Functional Equations: History, Applications and Theory

# Mathematics and Its Applications

# Functional Equations: History, Applications and Theory

edited by

J. Aczél

*Centre for Information Theory, University of Waterloo, Ontario, Canada*

D. Reidel Publishing Company

A MEMBER OF THE KLUWER ACADEMIC PUBLISHERS GROUP

Dordrecht / Boston / Lancaster

Library of Congress Cataloging in Publication Data

Main entry under title:

Functional equations: History, Applications and Theory

    (Mathematics and its applications)
    Includes index.
      1.   Functional equations–Addresses, essays, lectures.
I.  Aczél, J.  II.  Series: Mathematics and its applications (D. Reidel
Publishing Company)
QA431.F79          1984          515.7          83-24732
ISBN 1-4020-0329-3
Transferred to Digital Print 2001

Published by D. Reidel Publishing Company,
P.O. Box 17, 3300 AA Dordrecht, Holland

Sold and distributed in the U.S.A. and Canada
by Kluwer Academic Publishers,
190 Old Derby Street, Hingham, MA 02043, U.S.A.

In all other countries, sold and distributed
by Kluwer Academic Publishers Group,
P.O. Box 322, 3300 AH Dordrecht, Holland

# Table of Contents

# EDITOR'S PREFACE

Approach your problems from the right end and begin with the answers. Then one day, perhaps you will find the final question.

'The Hermit Clad in Crane Feathers' in R. van Gulik's The Chinese Maze Murders.

It isn't that they can't see the solution. It is that they can't see the problem.

G.K. Chesterton. The Scandal of Father Brown 'The Point of a Pin'.

Growing specialization and diversification have brought a host of monographs and textbooks on increasingly specialized topics. However, the "tree" of knowledge of mathematics and related fields does not grow only by putting forth new branches. It also happens, quite often in fact, that branches which were thought to be completely disparate are suddenly seen to be related.

Further, the kind and level of sophistication of mathematics applied in various sciences has changed drastically in recent years: measure theory is used (non-trivially) in regional and theoretical economics; algebraic geometry interacts with physics; the Minkowsky lemma, coding theory and the structure of water meet one another in packing and covering theory; quantum fields, crystal defects and mathematical programming profit from homotopy theory; Lie algebras are relevant to filtering; and prediction and electrical engineering can use Stein spaces. And in addition to this there are such new emerging subdisciplines as "completely integrable systems", "chaos, synergetics and large-scale order", which are almost impossible to fit into the existing classification schemes. They draw upon widely different sections of mathematics.

This program, Mathematics and Its Applications, is devoted to such (new) interrelations as exempla gratia:

- a central concept which plays an important role in several different mathematical and/or scientific specialized areas;
- new applications of the results and ideas from one area of scientific endeavor into another;
- influences which the results, problems and concepts of one field of enquiry have and have had on the development of another.

J. Aczél (ed.), Functional Equations: History, Applications and Theory, vii-ix.
© 1984 by D. Reidel Publishing Company.

The Mathematics and Its Applications programme tries to make available a careful selection of books which fit the philosophy outlined above. With such books, which are stimulating rather than definitive, intriguing rather than encyclopaedic, we hope to contribute something towards better communication among the practitioners in diversified fields.

The topic of the present volume in the MIA Series is functional equations, that is the search for functions which satisfy certain functional relationships such as, for example, $f(x + y) = f(x) + f(y)$ - one of the very simplest examples. They are remarkable in that they arise in most parts of mathematics. A number of relatively well known examples are Cauchy's equations (the one just written down), the functional equations for the Riemann zeta function (are there more solutions, a question studied by A. Weil), the equation for entropy, numerous equations in combinatorics, and quite recently for example the important Yang-Baxter equations in lattice statistical dynamics and quantum field theory which asks for $n^2 \times n^2$ matrix valued functions satisfying

$$(I \otimes R(u-v))(R(u) \otimes I)(I \otimes R(v)) = (R(v) \otimes I)(I \otimes R(u))(R(u-v) \otimes I)$$

(both sides are $n^3 \times n^3$ matrices). Still other examples arise in probability theory ($f((x^2 + y^2)^{1/2}) = f(x) f(y)$) in operator theory and in geometry (find all transformations $\mathbb{R}^n \to \mathbb{R}^n$ which take straight lines into straight lines).

In addition, there are functional equations which govern the behaviour of iterations of maps such as the Poincaré equation $F(az) = aF(z)(1-F(z))$ where $a$ is a parameter and the equation which is at the heart of various universality results (Feigenbaum theory, chaos theory), $G(n) = -\lambda^{-1} G(G(\lambda n))$, $\lambda = -G(1)$. And for that matter a formal group is an n-tuple of power series in 2n variables $F_1 (X_1, \cdots, X_n; Y_1, \cdots, Y_n)$, $\cdots, F_n (X_1, \cdots, X_n; Y_1, \cdots, Y_n)$ satisfying the functional equations $F(0; Y) = Y$, $F(X, 0) = X$, $F(F(X, Y), Z) = X$, $F(F(X, Y), Z) = F(X, F(Y, Z))$.

In turn, formal groups have, for example, application in algebraic geometry, algebraic number theory and algebraic topology. Thus functional equations seem to occur in virtually all parts of mathematics.

The subject does not fit well into any of the established mathematical specialisms, neither in terms of its problems, nor in terms of its results and techniques. Though there are several spectacular results such as Weierstrass' theorem in which kinds of functions admit some sort of polynomial addition formula, a general, universal powerful technique for dealing with functional equations does not yet exist and thus the material in this book poses a considerable challenge to the mathematical community. This fits in well with the philosophy behind the Mathematics and Its Applications book series. Also a glance at the Table of Contents will convince the reader that the subject impinges on even more specialisms in mathematics then are mentioned above, making this a truly multispecialistic volume, another aim of the MIA programme.

The unreasonable effectiveness of mathematics in science ....

Eugene Wigner

Well, if you knows of a better 'ole, go to it.

Bruce Bairnsfather

What is now proved was once only imagined.

William Blake

As long as algebra and geometry proceeded along separate paths, their advance was slow and their applications limited.
     But when these sciences joined company they drew from each other fresh vitality and thenceforward marched on at a rapid pace towards perfection.

Joseph Louis Lagrange

Amsterdam, October 1983                     Michiel Hazewinkel

# Essays

J. Aczél

## On history, applications and theory of functional equations

**1.** Functional equations have much in common with the *axiomatic method:* they define their objects (functions) implicitly by their properties (equations) rather than by direct definitions (explicit formulae).

Indeed, while functional equations "in a single variable"[1] in the form of recurrent sequences go back to antiquity (Archimedes, for instance; but they still bring intriguing, useful and picturesque new results; cf. paper[2] no. 10 by R. Thibault and essay no. 6 by Sklar), Oresme who seems to have been the first mathematician to use functional equations "in several variables"[1], did so exactly for the purpose of such an indirect definition of linear functions. In fact, he *defined* these "uniformly deformed qualities" by what amounts to the functional equation

$$\frac{f(x_1)-f(x_2)}{f(x_2)-f(x_3)} = \frac{x_1-x_2}{x_2-x_3} \quad \text{for all} \quad x_1,x_2,x_3 \text{ with } x_1 > x_2 > x_3. \quad (1)$$

The relevant passages in Oresme c. 1352[2] say "Qualitas vero uniformiter difformis est cuius omnium trium punctorum proportio distantie inter primum et secundum ad distantiam inter secundum et tertium est sicut proportio excessus primi supra secundum ad excessum secundi supra tertium in intensione" ... "Omnis autem qualitas se habens alio modo a predicto dicitur difformiter difformis et potest describi negative, scilicet qualitas que non est in omnibus partibus subiecti equaliter intensa nec omnium trium punctorum ipsius proportio excessus primi supra secundum ad excessum secundi supra tertium est sicut proportio distantiarum eorum". ("A uniformly deformed quality is one for which, given any three points, the ratio of the distance between the first and second point to that between the second and third equals the ratio of the excess in

3

J. Aczél (ed.), *Functional Equations: History, Applications and Theory*, 3-12.

intensity of the first over the second to that of the second over the third" ... "Every quality which behaves in any way other than those previously described is said to be deformedly deformed. It can be described negatively as a quality which is neither everywhere equally intense nor for which, given any three points, the ratio of the excess of the first over the second to the excess of the second over the third equals the ratio of their distances.") The "previously described qualities" were the "uniform" (constant) ones and those just defined as "uniformly deformed" (linear).

If we choose $x_2$ half way between $x_1$ and $x_3$, we get what is now called "Jensen's functional equation" (cf. Aczél 1966 or Aczél-Dhombres 1985[3] for this and other points, in particular where no specific references are made)

$$f(\frac{x_1+x_3}{2}) = \frac{f(x_1)+f(x_3)}{2} . \tag{2}$$

Several other functions were subsequently defined by functional equations, like the trigonometric functions (by Tannery 1886), the elliptic functions (by Weierstrass, cf. Rausenberger 1884), the gamma function (by Artin 1931) etc. (cf. also essay no. 5 by L. Reich).

**2.** While we have first emphasized the relations of functional equations to the axiomatic method, this does not say by any means that they are not useful for *applications*. Quite the opposite: it happens rather often in the natural, social and behavioral sciences (as well as in mathematics itself) that one or more properties of an unknown function (formula, "law") describing a process are known and can be written in the form of functional equations. If we know all solutions of these equations, that is, all functions in a certain class (determined usually by regularity, say boundedness or differentiability, conditions - the fewer the better - and by initial and/or boundary conditions) which satisfy those equations, then we obtain the complete description of that process. This occurs quite regularly since, as Galileo said "the laws of nature are written in mathematical language".

A first spectacular example of this goes back precisely to Galileo and further. It was Oresme, even earlier (Oresme c. 1347), who verbally described for the quadratic function a functional equation which (up to a shift) can be written in modern notation as

$$\frac{s[(n+1)t]-s(nt)}{s(nt)-s[(n-1)t]} = \frac{2n+1}{2n} \tag{3}$$

for all positive $t$ and all positive integers $n$. The same equation with the same solutions (with just about the same description) reappears almost three centuries later with Galileo (1638). While neither Oresme nor Galileo proved that $s(t)=at^2$ is the *only* solution (with the initial condition $s(0)=0$ which Galileo has explicitly stated; continuity or monotony would be a convenient regularity supposition), Galileo was so convinced of this, that it was (3) which he verified in his experiments in order to show that the fall of bodies follows the quadratic law.

Later, in the course of the 18th, 19th and 20th centuries the parallelogram law of composition of forces was reduced to the equations

$$f(x+y) = f(x)+f(y) \tag{4}$$

$(x,y \geq 0)$ and

$$g(x+y)+g(x-y) = 2g(x)g(y) \quad (0 \leq y \leq x \leq \frac{\pi}{2}) \tag{5}$$

(which came to be called Cauchy's and d'Alembert's equation, respectively) and solved under satisfactorily weak conditions. Also, in mathematics, Newton's binomial law

$$(1+x)^\alpha = \sum_{n=0}^{\infty} \binom{\alpha}{n} x^n$$

was proved by Euler, Lacroix, Cauchy and Abel using (4) and the further Cauchy equations

$$f(x+y) = f(x)f(y) \tag{6}$$

$$f(xy) = f(x)+f(y) \quad (x,y>0) \tag{7}$$

(see essay no. 3 by J. Dhombres). The fourth Cauchy equation is

$$f(xy) = f(x)f(y). \tag{8}$$

The functional equation (7) served earlier (Briggs 1624) to *construct* the logarithmic function, and when Saint Vincent (1647) noticed that the area beneath the hyperbola satisfies the same equation, de Sarasa (1649) soon deduced that it *has to be* a logarithm (in modern notation $\int_1^x \frac{1}{t}\,dt = \ln x$).

Applications of functional equations, both inside and outside mathematics, abound also nowadays. We mention just applications to iteration, dynamical systems (cf. essay no. 6 and paper no. 12 by A. Sklar, paper no. 11 by R.L. Clerc and C. Hartman, and essay no. 5 by L. Reich) to probabilistic metric spaces (paper no. 6 by Cl. Alsina), to group theory, mean values, physics (cf. paper no. 1 by J. Dhombres) and, in particular, to economics, decision and information theory (paper no. 7 by B.R. Ebanks, paper no. 8 by J. Aczél, essay no. 4 and paper no. 9 by W. Gehrig). These and other applications of functional equations and their closely related theory show the double aspects of esthetics and usefulness of functional equations (see essay no. 2 by Cl. Alsina).

**3.** This theory has developed through the solution of equations like (4) under ever *decreasing regularity suppositions* (which is also important for applications) to *solution methods and theorems* for wide *classes* of functional equations. Abel (1827) introduced the first such method: reduction to differential equations, while noticing at the same time the remarkable fact (unparalelled for other equations) that *one* functional equation can determine *several* unknown functions. We quote from this paper: "In der That lassen sich durch wiederholte Differentiationen nach den beiden veränderlichen Grossen, so viel Gleichungen finden, als nöthig sind, um beliebige Functionen zu eliminieren, so dass man zu einer Gleichung gelangt, welche nur noch eine dieser Functionen enthält und welche im Allgemeinen eine Differential-Gleichung von irgend einer Ordnung sein wird. Man kann also im Allgemeinen alle die Functionen vermittelst einer einzigen Gleichung finden. Daraus folgt, dass eine solche Gleichung nur selten möglich sein wird. Denn da die Form einer beliebigen Function die in der gegebenen Bedingungs-Gleichung vorkommt, vermöge der Gleichung selbst, von den Formen der andern abhängig sein soll, so ist offenbar, dass man im Allgemeinen keine dieser Functionen als gegeben annehmen kann." ("Actually, as many equations can be found by repeated differentiations with respect to the two independent variables as are necessary to eliminate arbitrary functions. In this manner, an equation is obtained which contains only one of these functions and which will generally be a differential equation of some order. Thus it is generally possible to find all the functions by means of a single equation. From this it follows that such an equation can exist only rarely. Indeed, since the form of an arbitrary function appearing in the given conditional equation, by virture of the equation itself, is to be dependent on the forms of the others, it is obvious that, in general, one cannot assume any of these functions to be given.")

On the other hand, Hilbert (1900) wrote in connection with the fifth of his famous problems "Uberhaupt werden wir auf das weite und nicht uninteressante Feld der Funktionalgleichungen geführt, die bisher meist nur unter Voraussetzung der Differenzierbarkeit der auftretenden Funktionen untersucht worden sind. Insbesondere die von Abel mit so vielem Scharfsinn behandelten Funktionalgleichungen... und andere in der Literatur vorkommenden Gleichungen weisen an sich nichts auf, was zur Forderung der Differenzierbarkeit der auftretenden Funktionen zwingt..." ("Specifically, we come to the broad and not uninteresting field of functional equations, hitherto largely investigated by assuming differentiability of the occurring functions. Equations treated in the literature, particularly the functional equations treated by Abel with such incisiveness, show no intrinsic characteristics that require the assumption of differentiability of the occurring functions...").

Such elimination or reduction of regularity conditions was indeed achieved later by different methods, one of the most important of which uses the distributions of L. Schwartz (see paper no. 3 by A. Tsutsumi and Sh. Haruki). - Even more general methods for solving equations for *multiplace* functions are presented by A. Krapež (essay no. 7 and paper no. 2) on very general algebraic structures.

**4.** The *domains of the equations* and ranges of the unknown functions can make quite a difference. For instance, for (6), if the domain is $\{(x,y)|x>0,y>0\}$ then every solution has a (unique) *extension* which satisfies (6) for all real $x,y$. If the domain is $\{(x,y)|x\geq0,y\geq0\}$, then $f(x)=1$ for $x=0$ and 0 for $x>0$ is a solution, but has no such extension at all (for extensions and conditional equations, cf. paper no. 1 by J. Dhombres).

Also, complex functions of a complex variable can show remarkably different behavior, as far as functional equations are concerned, than real functions of a real variable. A notable example is the fact that, while it is easy to see, without any regularity suppositions, that $f(t)=t$ and $f(t)=0$ are the only real solutions of *both* (4) and (8), that is, the only endomorphisms of reals into reals, this is not so anymore in the complex field, even if the additional solution $f(t)=\overline{t}$ (conjugate) is taken into consideration. This was shown by counter-examples constructed by Segre (1947) and Kestelman (1951). It is interesting to note that Lebesgue (1907) used a previous (incomplete) construction of such a counterexample (based on the well ordering of the continuum) to explain the difference between the axiomatic ("idéaliste") and intuitionist ("empiriste") points of view: "Cela prouve pour les Idéalistes l'existence d'une infinité de solutions différentes pour votre problème. Le raisonnement qui précède n'a au contraire guère de valeur pour un Empiriste, tout au plus, et seulement pour ceux des Empiristes qui admettent qu'on peut utiliser dans une définition des conditions telles qu'on ne sache ni vérifier qu'elles sont remplies, ni vérifier qu'elles ne le sont pas - montre-t-il si l'on savait bien ordonner le continu, on saurait nommer des solutions de votre problème autres que les solutions connues". ("This proves for the 'idealists' the existence of infinitely many different solutions of your problem. On the other hand, the preceding reasoning is of no value at all for an 'empiricist'. At most it would mean the following - and only to such empiricists who allow in a definition the use of conditions about which one can verify neither that they are satisfied, nor that they are not: If the continuum could be well ordered, then one could find solutions to your problem different from the previously known solutions.")

Another way in which complex functional equations differ from real ones is that regularity (differentiability) conditions and, in particular, the supposition that the solution be meromorphic or even entire, is much stronger than for reals since one can use the whole arsenal of complex function theory (cf. paper no. 4 by H. Haruki). Conversely, by omitting regularity conditions, even the classical problems of (linear and algebraic) independence and dependence of solutions of (4) or (6) can be looked upon from a different point of view (see paper no. 5 by L. Reich and J. Schwaiger).

Since the innovative books of Hille (Hille 1948, Hille-Philips 1957), the theory of functional equations for operators has flourished. Some of these equations (defining Reynolds and related operators) again go back to very practical hydrodynamical problems, among others (cf. essay no. 3 and paper no. 1 by J. Dhombres).

**5.** Of course, this short survey cannot do full justice to any of the three subjects in its title: history, applications and theory of functional equations, and to the vast perspectives this subject has. Still I hope that it has given some feeling of the abundance of ideas in this discipline and that it is of some use as introduction to this book. Fortunately, the other essays and papers in the volume show this wealth of ideas and applications even more.

This book originated from talks and from a panel discussion at the Functional Equations section of the Second World Conference on Mathematics at the Service of Man(kind). It contains selected talks[4] (as 'papers') and all panel contributions (as 'essays'). However, almost all articles (in particular those which have appeared or are to appear in different form elsewhere) were thoroughly rewritten, incorporating more recent results, among others.

Thanks are due to Professors E. Trillas and Cl. Alsina for organizing the congress, to Professor A. Sklar for helping select the papers, to Professor H.J. Skala for suggesting that this be published, to the Reidel Publishing Company for publishing it in this nice form, to Mrs. Linda Gregory for the enormous work of typesetting the whole book on a computer-wordprocessor, to Dr. June Lester for help with the index and proofreading and for improvements in style and language, to Stephen Birkett, also for help with the index and, last but not least, to the authors whose excellent work made the whole enterprise worthwhile.

Centre for Information Theory
University of Waterloo
Waterloo, Ont., Canada
N2L 3G1

## Notes

1.  The difference between equations in single or several variables is not whether the unknown function is single or multiplace, but whether there is one variable in the equation, like in $f(x+1)=F[f(x)]$ or several like in (1) and (2).

2.  Essays and papers appearing in this volume are quoted by their number and authors, other references are given with the author's name and year of publication and refer to the bibliography at the end of the present essay.

3.  The historical parts of the present essay rely heavily on Chapter 21 of that book (Aczél-Dhombres 1985).

4.  Papers no. 8, 1 and 2 were originally the main lecture of the congress, the main talk and the long talk in the functional equations topic, respectively; but these distinctions have faded with time.

## Bibliography

[1]   Abel, N.H.: 1827, Uber die Functionen, die der Gleichung $\phi x+\phi y=\phi(xfy+yfx)$ genug thun. *J. Reine Angew. Math. 2*, 386-394 (Oeuvres completes, Vol. 1, pp. 389-398, Christiania, 1881).

[2]   Aczél, J.: 1966, *Lectures on Functional Equations and Their Applications*. Academic Press, New York-London.

[3] Aczél, J. and J. Dhombres: 1985, *Functional Equations Containing Several Variables*. Addison-Wesley, Reading, Mass.

[4] Artin, E.: 1931, *Einführung in die Theorie der Gammafunktion*. Teubner, Leipzig.

[5] Briggs, H.: 1624, *Arithmetica Logarithmica*. London.

[6] Galileo, G.: 1638, *Discorsi e dimostrazioni matematiche intorno a due nuove scienze*. Leyden (Opere, Vol. VIII, pp. 209-210. Barbèra, Firenze, 1968).

[7] Hilbert, D.: 1900, Mathematische Probleme (Vortrag, gehalten auf dem internationalen Mathematiker-Congress zu Paris, 1900). *Nachr. Ges. Wiss. Göttingen* 253-297 (Gesammelte Abhandl., Vol. III, pp. 290-329, Berlin, 1935).

[8] Hille, E.: 1948, *Functional Analysis and Semi-groups*. (Colloq. Publ. Amer. Math. Soc., Vol. 31), New York.

[9] Hille, E. and R.S. Philips: 1957, *Functional Analysis and Semi-groups*. (Colloq. Publ. Amer. Math. Soc., Vol. 31), Providence, RI.

[10] Kestelman, H.: 1951, Automorphisms in the field of complex numbers. *Proc. London Math. Soc. (2) 53*, 1-12.

[11] Lebesgue, H.: 1907, Sur les transformations ponctuelles transformant les plans en plans, qu'on peut définir par des procédés analytiques. *Atti Accad. Sci. Torino 42*, 532-539.

[12] Oresme, N.: c. 1347, *Questiones super geometriam Euclidis* (Manuscript). Paris (Ed. H.L.L. Busard; E.J. Brill, Leiden, 1961).

[13] Oresme, N.: c. 1352, *Tractatus de configurationibus qualitatum et motuum* (Manuscript). Paris (Ed. transl. and comm. M. Clagett; Univ. of Wisconsin Press, Madison-Milwaukee-London, 1968).

[14] Rausenberger, O.: 1884, *Lehrbuch der Theorie der periodischen Funktionen*. Teubner, Leipzig.

[15] de Saint-Vincent, G.: 1647, *Opus geometricum quadraturae circuli et sectionem coni. (Problema Austriacum. Plus ultra quadratura circuli)*. Antverpiae.

[16] de Sarasa, A.: 1649, *Solutio problematis a R.P. Marino Mersenno minimo propositi*. Antverpiae.

[17] Segre, B.: 1947, Gli automorfismi del corpo complesso ed un problema di Corrado Segre. *Atti Accad. Naz. Lincei Rend. (8) 3*, 414-420.

[18] Tannery, J.: 1886, *Introduction à la théorie des fonctions d'une variable.* §96. Paris.

Cl. Alsina

# The esthetics and usefulness of functional equations

The theory of functional equations is fascinating because of its intrinsic mathematical beauty as well as its applications. Let us talk about beauty. Of course, functional equations share with Mathematics a sense of beauty common to all branches of this science (and art!) but they have also some characteristic aspects of their own. In this field one deals with mathematical identities where the solutions strongly depend upon the domains and the regularity assumptions required for the unknowns. Proofs are usually clear, clean, short; elegant arguments come up. Sometimes the equations give you just a little information, but, by using the powerful methods that the theory provides, you can say quite a lot about the general solutions. You look for existence and uniqueness. You look for characterizations. You get a lot from little. That is nice.

You may realize that in fact few mathematicians were born functional equationists in their mathematical life. They became involved in the field initially through some motivating problems. What fascinated them? A feeling of deep mathematical searching for beauty is certainly behind that process. And beauty in Mathematics means, among other things, a certain generality, a certain depth - the possibility of significant ideas ... .

The theory of functional equations is a growing branch of mathematics which has contributed greatly to the development of strong tools in today's Mathematics and, conversely, many mathematical ideas in several fields have become essential to the foundations of functional equations. In this way the theory has acquired its own personality. In the words of Whitehead "it is the large generalization, limited by a happy particularity, which is the fruitful conception."

J. Aczél (ed.), Functional Equations: History, Applications and Theory, 13-15.

And now let us turn our attention to applications.

My personal feeling is that in our day there are basic real-life problems which cannot be handled by "classical" Mathematics. They require a new machinery to be solved, or at least, to be formulated precisely. Let us face some questions: What is information? How can we measure the different kinds of information we get? Which parameters can be considered in order to measure human health? What logic is behind our brain? What is vagueness? How can we synthesize judgements? Can we analyse indistinguishability relations? Which measures are essential in economics? How do we learn? ... etc. A long list of fascinating questions is waiting for precise answers and new fields are growing up looking for new models, new arguments: Information Theory, Software Sciences, Cognitive Science, Inductive Logic, Multivalued Logic, Psychometry, Fuzzy Sets Theory, Mathematical Economics, Analytic Hierarchy Processes, Cluster Analysis, Artificial Intelligence, Probabilistic Metrics, Iteration Theory, etc. These fields use classical mathematics but they need to go further. In all of them there is a need to measure, to synthesize sets of data, ratios, frequencies ... and functional equations arise naturally. The theory of functional equations has contributed strongly to the attack and solution of many problems suggested by the above mentioned theories and other fields. Many "new" applied problems and theories have motivated functional equationists to develop new approaches and new methods. From my own experience, I can say that working in Barcelona in the interdisciplinary seminar of Professor Trillas, I could help people many times to formulate and solve problems using my modest knowledge of functional equations and inequalities and, reciprocally, I did some work solving equations by starting from some applied problems which arose from that seminar. For example, in this "functional equation service", as my colleagues informally call our collaboration, we became interested in the study of general logical connectives and negations, i.e., in the construction of some classes of De Morgan algebras which could be of interest in pattern recognition by computers and robots. In order to face this problem, the key was to consider, instead of the classical $(Min, Max, 1-j)$, the triples $(T, T^*, n)$, where $T$ is a suitable topological semigroup on the unit interval. But for solving concrete problems and finding new characterizations of logical operators the basic tool was the representation theorem of Aczél-Ling for the solutions of the associativity functional equation. In this framework the classical structure of Boolean algebra was a particular case of that of a De Morgan algebra in the case where distributivity functional equations were assumed. We have produced a large number of results (and papers) in this direction. Now the people in the seminar who raised these questions are using our results in their applied research.

To end this contribution let me recall the famous words of Hardy: "A mathematician, like a painter or a poet, is a maker of patterns. If his patterns are more permanent than theirs, it is because they are made with ideas". Functional equationists have developed beautiful and powerful ideas by building a large number of methods for solving their equations. On the initiative of Professor János Aczél, the theory has become an important matter of our culture.

I really don't know if Mathematics is "at the service of man" but certainly "men and women" may enjoy and apply Mathematics - if they know at least a little bit of functional equations.

Dep. Mathemàtiques i Estadística
(E.T.S.A.B.) Univ. Politècnica de Barcelona
Diagonal 649, Barcelona 28,
Spain

J.G. Dhombres

## On the historical role of functional equations

The best initial approach when confronted with the task of justifying or explaining activities in any branch of science is to look back, that is, to look at the development of this branch from a historical perspective. This supposes that our predecessors were not more stupid than we are now, a reasonable assumption.

A second approach has a historical connotation too, but with a far shorter range: it is a personal history which describes the way we have become involved with the subject.

A third approach concerns itself more with the present and describes what the subject looks like, putting the emphasis on recent achievements, on applications to other domains or on the enrichment coming from other subjects. Such an attitude naturally leads to a perspective: what to expect, what is worth solving, which projects can reasonably be undertaken.

For functional equations I shall not attempt to take the last attitude as other well-informed authors in this volume have concentrated on it. I shall try to elaborate on the first aspect, and to say a few words about the second. Unfortunately, the history of functional equations still waits for an author [1]. But every mathematician is aware of how this field was developed in the work of authors from the second part of the eighteenth century and throughout all the nineteenth century. Names like d'Alembert, Euler, Cauchy, Abel, and Riemann, to quote only some among the most famous, are all associated with important and deep research on functional equations. A look at the long bibliography in the book of J. Aczél [2] also testifies to the continuous and developing attraction of the field of functional equations during our century. I should like to elaborate first on the work of Cauchy.

*J. Aczél (ed.), Functional Equations: History, Applications and Theory, 17-31.*
© *1984 by D. Reidel Publishing Company.*

## 1. Cauchy's role in the development of functional equations

It is generally acknowledged that Cauchy brought rigor to analysis in the sense that he was able to initiate a standard way to formulate the usual proofs in analysis via a standardized use of limits, of $\epsilon$ and $\delta$, and of inequalities, even if we are aware of his systematic mistake about uniform properties. Typical examples are his proofs of the intermediate value theorem, the fundamental theorem of calculus, the definition of the definite integral, the Cauchy condition using

$$\lim \sqrt[n]{u_n}$$

for the convergence of series with nonnegative terms etc. Some of these innovative methods appeared in a textbook written when Cauchy (1789-1857) was under thirty, although the book, known as the *Cours d'analyse algébrique de l'Ecole Royale Polytechnique*, was published first in 1821. The introductory address of the book is famous and I shall just quote a passage which emphasizes the need for a rigorous treatment of analysis.

"Quant aux méthodes, j'ai cherché à leur donner toute la rigueur qu'on exige en géométrie, de manière à ne jamais recourir aux raisons tirées de la généralité de l'algèbre. Les raisons de cette espèce, quoique assez communément admises, surtout dans le passage des séries convergentes aux séries divergentes, et des quantités réelles aux expressions imaginaires, ne peuvent être considérées, ce me semble, que comme des inductions propres à faire pressentir quelquefois la vérité, mais qui s'accordent peu avec l'exactitude si vantée des sciences mathématiques." [See note [3] for a translation.]

In this book of Cauchy we find also the first systematic treatment of some basic functional equations, the now classical Cauchy equations like

$$f(x+y)=f(x)+f(y) \tag{1}$$

or

$$f(x+y)=f(x)f(y) \tag{2}$$

and the d'Alembert functional equation

$$f(x+y)+f(x-y)=2f(x)f(y). \tag{3}$$

What purpose do they serve in this book? A *first reason* for Cauchy to include them, was quite clearly to prove how powerful and practical are some rather abstract and original concepts introduced by him at the beginning of his book: the concept of limit, quantified by the use of $\delta$ and $\epsilon$, and the concept of a continuous function. Functional equations serve as proofs of his "savoir-faire". The word "limit" had been frequently used before Cauchy and d'Alembert emphasized it as the basic tool for Calculus. However, the superiority of Cauchy's limit concept lies mainly in the practical and systematical use he made of it. Regarding the word "continuous", Cauchy, in his cautious way, took it over from Euler, but completely modified its meaning and gave an operational definition [4]. Cauchy's way of proving that his definition of continuity did not belong to the metaphysical limbo of inefficient definitions, was precisely to find all continuous solutions of (1), (2) and (3) among numerical functions defined for one "real" variable. And he was indeed the first to do so without stronger regularity assumptions. More important is the fact that he emphasized the need for some regularity assumptions at all.

Our appreciation of Cauchy's achievement and savoir-faire grows if we compare it to earlier attempts, or even attempts in Cauchy's time, where no regularity assumption was made for the solutions of (1), (2) or (3) but such regularity seemed to be implicitly included in the very concept of a function. The overenthusiastic use of functional equations has also led occasionally to serious mistakes. That was the case for A.M. Legendre in various notes of his *Eléments de Géométrie* (first published in Paris in 1794 and quite often reissued during the nineteenth century). Legendre pretended for example to get a "functional" proof of Euclid's axiom about parallels (by looking at the sum of the three angles of a triangle) or to get a "functional" proof for the formula of the area of the circle ($S = kR^2$) avoiding integrals or Eudoxus' method of exhaustion. Almost nobody was convinced that Legendre's proofs were correct, and the outcome was a serious handicap to the development of the systematic use of functions outside analysis. Another example can be seen in Gauss [5].

The lack of a clear idea of which kind of regularity was hidden behind the definition of some given function obliged the more cautious authors to make detours. A good example can be seen in

an influential treatise on analysis first published in 1797 by S.F.
Lacroix (1765-1843). This author in his *Traité de calcul différentiel
et de calcul intégral* [6] wanted to prove Newton's binomial
theorem and wrote a priori

$$(1+x)^{\alpha} = 1+f(\alpha)x+g(\alpha)x^2+\dots \qquad (4)$$

He showed by induction that $f(\alpha)$ determines the other functions
$g(\alpha)$ etc. Clearly $f$ satisfies the functional equation

$$f(\alpha+\beta) = f(\alpha)+f(\beta). \qquad (5)$$

As $f(1)=1$, Lacroix deduced that $f(\alpha)=\alpha$ for rational $\alpha$. It seems
that he would have liked very much to deduce this result right
away for all values $\alpha$. Instead he provided a lengthly proof using a
very indirect approach making use also of the functional equation of
the logarithmic function. On the other hand, Cauchy cleared the
way for a functional analysis of this problem by his clever handling
of functional equations.

A *second reason* for Cauchy to include functional equations in
his    Course    goes    deeper.    Lacroix's    abortive    attempt
notwithstanding, functional equations formed decisive steps towards
the proof of Newton's binomial theorem in Cauchy's book, i.e., of

$$(1+x)^{\alpha} = 1+\alpha x+\frac{\alpha(\alpha-1)}{2!}x^2+\dots$$
$$+\frac{\alpha(\alpha-1)\dots(\alpha-n-1)}{n!}x^n+\dots, \qquad (6)$$

where $\alpha$ is any real number and $x$ a complex number whose
modulus is less than one. Because of our familiarity with the proof
via the Taylor expansion, we have forgotten how important, even
crucial, the result (6) was for analysts of the eighteenth century.
The problem was to obtain (6) for noninteger values of $\alpha$ and even
for complex values. Previously a proof for $\alpha=1/2$ was formally
extended to all real and complex values of $\alpha$. Euler tried many
times [7] using derivatives, to get a correct and convincing proof,
but he noticed that there was a vicious circle, as (6) was used at the
very beginning of the calculus precisely to compute derivatives of
the power functions. We touch here the most interesting aspect of
Cauchy's endeavour while writing his course. Not only did he try
to be as rigorous as the mathematics of his time permitted, but he

organized his material in a linear way, let us say along the
*euclidean pattern.* Recall that one of the achievements of the
Elements was the idea of putting things one after another in a
specific order. Some few basic concepts were introduced by Cauchy
first (functions, limits, continuity ...) in the not clearly defined
setting of measurable quantities (what we call nowadays real
numbers but which were not clarified until Dedekind and Cantor
did so around 1870; see [8].) From there on, with a rare sense of
economy, Cauchy tried to obtain as much as he could without
introducing the derivative, which was to be the cornerstone of
another part of analysis, the infinitesimal analysis.

What really appears in the course when considered in its
entirety, is an elegant and systematic *construction* of a large bulk of
mathematical knowledge, based on the technique of convergence,
which is fundamental for (6). *It is exactly for this construction that
functional equations play an important role.* This aspect deserves to
be explained. Let us quickly recall Cauchy's method, starting at
the end, that is, the proof of (6). We define with Cauchy

$$f(\alpha) = 1+\alpha x+\frac{\alpha(\alpha-1)}{2!}x^2+...$$
$$+\frac{\alpha(\alpha-1)...(\alpha-n+1)}{n!}x^n+... \,. \qquad (7)$$

A whole chapter of Cauchy's book was devoted to convergence
criteria for such power series, both in the real and in the complex
case. With their aid Cauchy deduced the convergence of this series
for all real $\alpha$ if $|x|<1$. Then Cauchy used a theorem asserting that
continuity is invariant under limit processes in order to get the
continuity of $f$. In this generality, Cauchy's theorem is false
because the uniformity of convergence was not supposed [9]. Then,
using a technique he specially devised for multiplying two power
series and some combinatorics, Cauchy got the functional equation

$$f(\alpha+\beta) = f(\alpha)f(\beta). \qquad (8)$$

He had prepared his way well because in an earlier chapter he
had already solved (8) for real valued continuous functions $f$. A
lengthy but clever proof had also been provided for the case of
complex-valued functions. This proof is necessary to get (6). It is

not just a mathematical curiosity. Finally,

$$f(\alpha) = A^\alpha$$

and so (6) is proved at least for all real $\alpha$. Almost all results from Cauchy's textbook were used in this proof, and functional equations played the key role around which technical methods were developed.

Five years later, N.H. Abel (1802-1829) went further, by allowing $\alpha$ to be complex, in a beautiful paper in which he further increased the number of functional equations involved in the proof [10]. This was the first convincing proof of the binomial theorem in this generality!

I took the risk of talking too long about Cauchy just to make clear that functional equations appeared in his work as useful tools to impress upon analysts the need for a rigorous treatment of analysis. This task always remains an objective in teaching, and in this way we understand the *pedagogical importance of functional equations* better. However, with Cantor's theory of sets and topology, for example, we have now other examples of mathematical construction, and we often forget to mention the method of functional equations explicitly.

However, the importance of functional equations is far more than a pedagogical gadget. Mathematicians always create newer concepts or objects, and we are confronted with the task of characterizing the new objects by a minimal number of properties, so that we get a better grasp of the essence of the new objects in relation with others. Quite often functional equations help us clarify various properties and select those we need. Let us take a second example from the history of mathematics.

## 2. The functional method: the role of d'Alembert

A second illuminating example of the role of functional equations is seen in physics with the advent of what we will call the *functional method,* a variant of (relative) axiomatization. By observation and "common sense guided by reason" (Descartes), the mathematization of a physical process may have resulted in some kind of 'laws' (formulas), giving the explicit functions describing the process. The functional method consists in finding the functions as a *necessary* consequence of some general laws which are basic to the physical situation being described. Functional equations in an extended sense appear naturally as the way of determining a function. This is precisely how the first differential equations and the first partial differential equations were introduced. But functional equations in a restricted sense were also introduced for the same purpose. A good example concerns the composition of two forces, that is the parallelogram rule, or in a more contemporary language, the problem of finding the resultant of two vectors.

Even though the clarification of what a force is took a long time, leading mathematicians like Newton tried to deduce the composition of forces from the experimentally observed composition of motions. In 1726, Daniel Bernoulli (1700-1782) rejected this experimental approach as suspect and unnecessary. His opinion about such a proof was severe: "Nihil in illa demonstratione ut falsam rejicio, sed quaedam ut obscura, quaedam ut non necessario vera" [11]. Then he proposed a geometrical approach, and functional equations were used. Later in the century various interesting attempts were made with different kinds of functional equations.

The most pertinent effort was made by J. d'Alembert (1717-1783) who introduced for this purpose the functional equation

$$f(x+y)+f(x-y) = f(x)f(y) \qquad (9)$$

in a short paper entirely devoted to the functional method [12].

D'Alembert was well prepared to find regular solutions for (9) since he had already introduced the functional equation

$$f(x+y)-f(x-y) = g(x)h(y), \qquad (10)$$

while dealing with vibrating strings [13]. In fact the left hand side of (10), when $x$ denotes the time and $y$ denotes the abscissa for a string, is the general form of the motion of a vibrating string. The right hand side indicates the separation of variables. S.D. Poisson (1781-1890) developed the study of (9), in the form (3), to prove the composition law of forces in his *Mécanique rationnelle* of 1805. Other authors like G. Monge (1746-1818) and P.S. Laplace (1749-

1827) also introduced functional equations for the same purpose. In 1876, Darboux reduced the problem of the parallelogram of forces to the solution of the first Cauchy functional equation (1) where this solution is supposed to be nondecreasing. We may mention in the same vein the study of bijective transformations of the real plane, mapping a straight line into a straight line, which Darboux called the fundamental theorem of projective geometry and where functional equations are essential. The failure to obtain similar results in the case of the complex plane led to a detailed study of nonmeasurable solutions of (1) with works of H. Lebesgue and C. Segre [14].

Naturally, the functional method applied to mathematics as well. J. Tannery in his *Introduction à la théorie des fonctions d'une variable* (Paris, 1886) used functional equations in order to introduce sin $x$ and cos $x$ a priori in analysis without any recourse to geometry. Instead of (3), he used a system of functional equations

$$\left.\begin{aligned}\phi(x+y) &= \phi(x)\phi(y)-\psi(x)\psi(y)\\\psi(x+y) &= \psi(x)\phi(y)+\psi(y)\phi(x)\end{aligned}\right\} \tag{11}$$

His ideas were similar to those of D. Bernoulli and related to the discovery of noneuclidean axiomatics and to the arithmetization of analysis. "Il y'a un intérêt philosophique évident à introduire dans l'analyse le moins possible de données expérimentales, et il importe par conséquent de donner des fonctions sin $x$ et cos $x$ une définition qui repose uniquement sur la notion de nombre et n'emprunte rien à l'idée d'espace" [15].

Another example of the same attitude can be observed in the modern treatment of the Γ-function by means of the equations

$$f(x+1) = xf(x) \tag{12}$$

or

$$\sqrt{\pi}f(2x) = (2^{2x}-1)f(x)f(x+\frac{1}{2}) \tag{13}$$

as can be seen in the very elegant booklet of E. Artin (1931, *Einführung in die Theorie der Gammafunktion),* where functional equations and their solutions within certain classes of regular functions are the essential ingredients [16].

One may think that the previous historical comments mainly prove the interest of functional equations of the Cauchy type in the past and that nowadays the interest has shifted to completely different kinds of equations. For the reasons we gave on the need for clarification and axiomatization, in addition to the reasons of pure curiosity and even of mathematical esthetics, many new functional equations have been examined in the last fifty years. However, it is astonishing how attractive the general theme of Cauchy's equations remained to many mathematicians. One reason was that a fascinating part of research was and is to prove that rather exotic functional equations, with or without regularity conditions, over more or less abstract structures, are in fact *equivalent* to Cauchy's equation. A second reason is the increasing number of problems from various origins leading to Cauchy's equations. We quote without further details or references, problems in relativity theory, in additive number theory, in gas dynamics, in the social and behavioural sciences, etc. There, the general setting of conditional Cauchy equations seems interesting and has already brought many results. This means that in many of these problems we restrict the validity of the functional equation by some conditions about the variables.

## 3. Reynolds operators and turbulent fluids

I would like to conclude this very short presentation on a personal note. I began to work on a problem which was raised by the engineer O. Reynolds at the end of the nineteenth century while studying turbulence of water in  pipes in a practical way. The Navier-Stokes partial differential equation governing the movement is nonlinear and, under many boundary conditions, difficult to solve. One may . expect to obtain simpler partial differential equations by performing some local averaging processes on the movement and possibly on the boundary conditions, and to obtain in this way explicit solution for the averaged movement. Various averaging processes were proposed, in particular with ergodic integrals of the form

$$\frac{1}{2T} \int_{-T}^{T} f(x)\,dx$$

(and limits of such integrals). An a priori study was made which led to functional equations for linear operators [17]. In particular the so-called *Reynolds operators* were discovered, i.e. linear operators $P{:}A{\rightarrow}A$, where $A$ is an algebra of functions, such that

$$P(f{\cdot}Pg + g{\cdot}Pf) = Pf{\cdot}Pg + P(Pf{\cdot}Pg)$$

for all $f,g$ in $A$.                                                    (14)

Related to these operators are the *averaging operators*, i.e. the operators satisfying

$$P(f{\cdot}Pg) = Pf{\cdot}Pg \text{ for all } f,g \text{ in } A.$$       (15)

It is through these operators that I became interested in functional equations. It is fascinating that this last functional equation (15) leads to a completely different and efficient approach to turbulence theory by way of probabilistic methods. In this setting, the operator $P$ is a mathematical expectation. In this way, results had already been obtained by the functional equations method, before probability theory had been mathematically developed by Kolmogorov and used for turbulence theory, and these results were later reinterpreted in probabilistic terms. Nowadays, the use of averaging operators (those satisfying (15)) is a promising theoretical approach to signal processing, for instance to recover the correct signal when it is locally distorted. This shows the interest of functional equations like (14) and (15) per se, not only in the case where we look for solutions in the class of linear operators. It is interesting, for instance, to explore their equivalence under certain conditions. Some interesting papers were published in this respect in the last twenty years [18] and linked to other results originating in queuing theory, where an important functional equation due to Baxter is similar to (14). Applications then spread to domains like the geometry of Banach spaces, multiplier theory in Fourier analysis, combinatorics, logic, Hilbert transforms, von Neumann algebras and group representation theory. These constitute a rich mathematical harvest from the practical origin of this theory in turbulence theory and therefore a good reason to pursue a systematic study of functional equations derived from (14) or (15), just as other applications and prospects of applications motivate the study of other functional equations.

Institut de Mathématiques
Université de Nantes
2 Chemin de la Houssinière
F-44072 Nantes Cedex
France

## Notes

1. There is a description of what happened in functional equations before 1910 in a paper by S. Pincherle in the *Encyclopédie des sciences mathématiques pures et appliquées* (Paris, 1912, Volume II, 1, II.). But very little is said about what Euler, Laplace or Monge did for the subject and nothing about the controversy during the eighteenth century concerning functions. In fact, it seems that a useful history of functional equations would have to study simultaneously the evolution of the concept of a function. About the history of this concept a very interesting paper was written by A.P. Youschkevitch, Arch. Hist. Exact. Sc. *16* no. 1 (1976-1977), pp. 37-85: The concept of function up to the middle of the 19th Century. In a book to appear in collaboration with J. Aczél, we include a historical chapter concerning at least the material covered in the book *(Functional equations in several variables,* Encyclopedia of Mathematics and its Applications, Addison-Wesley). See also the section 0.2 History, pp. 5-12 in [2].

2. J. Aczél, *Lectures on functional equations and their applications.* Academic Press, New York-London, 1966.

3. "As to the methods, I tried to give them the rigour required in geometry so that I should never have to take recourse to arguments brought in from the generalities of algebra. Such arguments, although used rather often, especially in going over from convergent to divergent series and from real quantities to imaginary expressions, seem to me just inductive inferences which sometimes let us guess the correct results, but they have little to do with the exactness which is so highly spoken of in the mathematical sciences."

4.  In a paper entitled "De usu functionum discontinuarum in analysi" published in 1767 (see Opera Omnia, Pars I, Vol. 23, p. 74), Euler made clear what he meant by a continuous curve. His definition required that all points of the continuous curve had to be determined by the same equation, as by a law: "Iam vero notissimum est, in Geometria sublimiori alias lineas curvas considerari non solere, nisi quarum natura certa quadam relatione inter coordinatas per quampiam aequationem expressa definiatur, ita ut omnia eius puncta per eandem aequationem tanquam legem determinentur. Quae lex cum principium continuitatis in se complecti censeatur, quippe qua omnes curvae partes ita vinculo arctissimo inter se cohaerent, ut nulla in illis mutatio salvo continuitatis nexu locum invenire possit ...". ("It is already well known that in advanced Geometry one usually considers only curves which are determined by some equation or law involving the coordinates. That law is supposed to conform with the principle of continuity, in as much as all consecutive parts of the curve are closely coherent so that there can be found no change breaking the bound of continuity at any point ...". Using Euler's definition, should the functional equation (1) be viewed as the "law" governing the idea of a "continuous" function, and therefore the relation $f(\alpha)=\alpha$ for rational $\alpha$ would imply $f(x)=x$ for all real $x$? To say the least, Euler's definition was not very useful for further understanding at least till the introduction of analytic extensions, which came much later.

5.  C.F. Gauss, Theoria motus corporum coelestium, liber II, sectio III, § 175-177, 1809 (Werke, Vol. VII, pp. 240-244, Leipzig, 1906).

6.  A second and enlarged edition, including Lacroix's *Traité des différences et des séries,* was published in 1810 and moulded the mathematical culture of many generations of analysts. The role of mathematical textbooks during that period has been discussed in J. Dhombres, *French mathematical textbooks from Euler to Cauchy* (to appear, 86 p.).

7. L. Euler, Demonstratio theorematis neutoniani de evolutione potestatum binomi pro casibus quibus exponentes non sunt numeri integri. Nova Comment. Acad. Sci. Petropol. 19 (1775), pp. 103-111. (Opera Omnia, Pars I, Vol. 15, pp. 207-216). There are various other attempts in Euler's works to prove Newton's binomial theorem. The vicious circle of the use of fluxions or derivatives was already noticed by Colson. (See page 308 of his book: *The method of fluxions by the inventor, Sir Isaac Newton*, 1736, London).

8. For a historical record of the evolution of ideas concerning "real numbers", see for example, J. Dhombres, *Nombre, mesure et continu: épistémologie et histoire*, Nathan, Coll. Cédic, Paris, 1978.

9. Cauchy "proved" that the limit of a sequence of continuous functions is continuous. Cauchy needed this for balancing the architecture of his Course as it linked the two basic concepts of analysis: continuity and limit. In 1826, Abel (see [10]) tried to correct Cauchy's error but he did not avoid all the pitfalls concerning these properties either. Gudermann was the first to sense the need for uniform convergence around 1837 and then the concept was used ten years later both by Stokes and Seidel. But their influence was weak and, surprisingly, it was Cauchy again who in 1853 gave the exact definition so that his theorem on the limit of continuous functions became true (Cauchy, Notes sur les séries convergentes...Oeuvres Complètes (1) t. 12, pp. 30-36). (See P. Dugac, *R. Dedekind et les fondements des mathématiques*, Vrin, Paris, 1976.)

10. N.H. Abel, Untersuchungen über die Reihe $1+(m/1)x+((m(m-1))/2)x^2+...$, Journal für reine und angew. Math. 1 (1826), pp. 311-339. See the original in French in Oeuvres Complètes, Vol. I, pp. 219-250, Christiania, 1881.

11. "I do not reject in this proof anything as false, but say that some points are obscure and not true by necessity." D. Bernoulli, Examen principorium mechanicae et demonstrationes geometricae de compositione et resolutione virium, Comm. Acad. Petrop. 1 (1726) pp. 126-142.

12. J. d'Alembert, Mémoires sur les principes de la mécanique. Mém. Académie Royale des Sciences 1769, pp. 278-286.

13. J. d'Alembert, Addition au mémoire sur la courbe que forme une corde tendue mise en vibration, Hist. Acad. Berlin 1750, pp. 355-360.

14. H. Lebesgue, Sur les transformations ponctuelles transformant les plans en plans qu'on peut définir par des procédés analytiques. Accad. Reale delle Scienze di Torino 42 (1906/07), pp. 219-226.

15. "There is an obvious philosophical interest in bringing into analysis as few experimental facts as possible. Therefore it is important to provide for the functions sin $x$ and cos $x$ a definition which only relies on the notion of numbers and has nothing to do with the notion of space".

16. This brings us to functional equations in a single variable. Such functional equations were studied from the 18th century on by mathematicians like C. Babbage. Babbage explained in 1813 (Phil. Trans. 105, pp. 389-423), "I am still inclined to think that the evolution of functional equations must be sought by methods peculiarly their own".

    The fascinating domain of functional equations in a single variable has been strongly developed in the last few years leading to a reinterpretation of older results and bringing new links to many mathematical theories. We may quote here, G. Targonski, *Topics in iteration theory*, Studia Math., Skript 6, Göttingen-Zürich, 1981, and of course, M. Kuczma, *Functional equations in a single variable,* Monografie Matematyczne, Tom. 46, Warszawa, 1968.

17. The paper by Reynolds appeared in 1895 in the Phil. Trans. Roy. Soc. London (pp. 123-164). Contributions came from M.L. Dubreil-Jacotin, J. Kampé de Fériet, G. Birkhoff. For references, see G.C. Rota, Reynolds operators, Proc. Symp. Appl. Math. Vol. 16 (1964), pp. 70-83, and J. Dhombres, *Sur les opérateurs multiplicativement liés,* Mémoire Soc. Math. France no. 27 (1971), 156p. For a nonprobabilistic approach to turbulence theory, see for example, J. Bass, *Les fonctions pseudo-aléatoires,* Mémorial des Sc. Math. Fasc 153, Gauthier-Villars, Paris, 1962.

18. For further references, see J. Dhombres, *Some aspects of functional equations,* Dept. of Math., Lecture Notes, Chulalongkorn University, 1979.

W. Gehrig

## Functional equation methods applied to economic problems: some examples

## 0. Introduction

During the last decade several fields of economic theory derived great benefit by the consequent application of functional equation methods. These applications refer to such different fields as

- the theory of neutral technical progress
- the theory of the price index
- marketing-theory
- the measurement of economic inequality
- oligopoly-theory.

This list is not at all complete. Many more disciplines could be listed. But, I think, this gives a typical sample and indicates how helpful functional equation methods are.

In what follows we give examples from all fields mentioned above. We will not go into details of the (in some cases rather lengthy) proofs.

## 1. Theory of neutral technical progress

In generalizing papers of Uzawa [1961] and Sato-Beckmann [1968], Gehrig [1976] considers production functions

$$F: \mathbf{R}_{++}^{n+1} \to \mathbf{R}, \quad (x_1,..,x_n;t) \to F(x_1,..,x_n;t)$$

($F$ twice continuously differentiable) \hfill (1.1)

($\mathbf{R}_{++} = \{x \mid x > 0\}$) and examines the effects of well-known forms of neutral technical progress on $F$. For example, let us assume that

J. Aczél (ed.), Functional Equations: History, Applications and Theory, 33-52.

$$\frac{F_i(\mathbf{x},t)}{F_j(\mathbf{x},t)} = \phi^{ij}\left(\frac{x_i}{x_j}\right) \quad \begin{cases} i<j \\ i=1,..,n-1 \\ j=2,..n \\ \phi^{ij}: \mathbf{R}_{++} \to \mathbf{R}_{++}. \end{cases} \qquad (1.2)$$

In (1.2) we postulate an invariant relationship between the marginal rate of substitution and the factor relation. This is called generalized Hicks neutral technical process (Hicks n.t.p. for short; see Hicks [1932]). Eq. (1.2) is equivalent to the existence of twice continuously differentiable functions $\Omega^{ij}$ and $H^{ij}$ such that

$$F(\mathbf{x},t)=\Omega^{ij}(x_1,..,x_{i-1},H^{ij}(x_i,x_j),x_{i+1},..,x_{j-1},x_{j+1},..,x_n,t), \qquad (1.3)$$

where $H^{ij}$ is linearly homogeneous. If we want to know whether a generalized Hicks n.t.p. is compatible for all pairs $(i,j)$, we have to consider the system of functional equations

$$\Omega^{12}[H^{12}(x_1,x_2),x_3,..,x_n;t]$$
$$= \cdots =\Omega^{1n}[H^{1n}(x_1,x_n),x_2,..,x_{n-1};t]$$
$$=\Omega^{i,i+1}[x_1,..,H^{i,i+1}(x_i,x_{i+1}),..,x_n;t]$$
$$= \cdots =\Omega^{in}[x_1,..,H^{in}(x_i,x_n),..,x_{n-1};t]$$
$$= \cdots =\Omega^{n-1,n}[x_1,...,x_{n-2},H^{n-1,n}(x_{n-1},x_n);t]. \qquad (1.4)$$

First we derive the form of the $H^{ij}$.

LEMMA (1.5). The explicit form of $H^{ij}$ is either

$$H^{ij}(x_i,x_j)=b_{ij}x_i^{S_{ij}}x_j^{1-S_{ij}}$$

$$(b_{ij}>0, 0<S_{ij}<1, \quad \text{constants}) \qquad (1.6)$$

or

$$H^{ij}(x_i,x_j)=(z_{ij}x_i^{\alpha}+w_{ij}x_j^{\alpha})^{1/\alpha}$$

$$(w_{ij},z_{ij}>0, \alpha \neq 0, \quad \text{constants}). \qquad (1.7)$$

According to (1.5), the $H^{ij}$ are either of $CD$ (see Cobb-Douglas [1928]) or of $ACMS$ (see Arrow-Chenery-Minhas-Solow [1961]) - type.

THEOREM (1.8). Let $F$ satisfy generalized Hicks n.t.p. for all pairs $(i,j)$. Then $F$ is either of the form

$$F(\mathbf{x},t) = \bar{B}[\beta x_1^{a_1}...x_n^{a_n};t]$$

$$\beta > 0,\ 0 < a_i < i\ (i=1,.,n)\quad a_1 + .. + a_n = 1,\ \text{constants} \qquad (1.9)$$

or of the form

$$F(\mathbf{x},t) = B[(g_1 x_1^{\alpha} + ... + g_n x_n^{\alpha})^{1/\alpha};t]$$

$$g_i > 0\ (i=1,...,n)\quad \alpha \neq 0,\ \text{constants}. \qquad (1.10)$$

In both cases, $F$ is a homothetic production function (see Shepard [1953, p. 40]). For details concerning the proofs of (1.5) and (1.8) and for further applications of functional equations to problems in the theory of neutral technical progress we refer to Gehrig [1976].

## 2. Price-index theory

In the axiomatic theory one considers the price index as a positive-valued function, which depends either

(i)   on the prices of $n$ commodities of a base year and of the current year or

(ii)  on both the prices and quantities of $n$ commodities of a base year and of the current year, and satisfies several properties, which are called *axioms*, because they are in a certain sense natural.

We denote by

$$\mathbf{p}^0 = (p_1^0,..,p_n^0)\in \mathbf{R}_{++}^n, \quad \mathbf{p} = (p_1,..,p_n)\in \mathbf{R}_{++}^n \qquad (2.1)$$

the vectors of the prices of $n$ commodities of a base year and of the current year, respectively.

The price index is a function

$$F:\begin{cases} \mathbf{R}_{++}^{2n}\to \mathbf{R}_{++} \\ (\mathbf{p}^0,\mathbf{p})\to F(\mathbf{p}^0,\mathbf{p}), \end{cases} \qquad (2.2)$$

depending on the vectors (2.1) and satisfying the following axioms (A1)-(A4).

## (A1) Monotonicity Axiom:
For all $\mathbf{p}^0,\bar{\mathbf{p}}^0,\mathbf{p},\bar{\mathbf{p}}\in \mathbf{R}_{++}^n$:

$$F(\mathbf{p}^0,\mathbf{p})>F(\mathbf{p}^0,\bar{\mathbf{p}})$$

if

$$\mathbf{p}>\bar{\mathbf{p}} \quad (\mathbf{p}>\bar{\mathbf{p}}: \Longleftrightarrow \mathbf{p}\geq\bar{\mathbf{p}} \text{ and } \mathbf{p}\neq\bar{\mathbf{p}})$$

and

$$F(\mathbf{p}^0,\mathbf{p})<F(\bar{\mathbf{p}}^0,\mathbf{p}) \quad \text{if} \quad \mathbf{p}^0>\bar{\mathbf{p}}^0.$$

## (A2) Linear Homogeneity Axiom:

$$F(\mathbf{p}^0,\lambda\mathbf{p})=\lambda F(\mathbf{p}^0,\mathbf{p}) \quad \forall\mathbf{p}^0,\mathbf{p}\in \mathbf{R}_{++}^n, \forall\lambda\in \mathbf{R}_{++}.$$

## (A3) Identity Axiom:

$$F(\mathbf{p}^0,\mathbf{p}^0) = 1 \quad \forall\mathbf{p}^0\in \mathbf{R}_{++}^n.$$

## (A4) Dimensionality Axiom:

$$F(\lambda \mathbf{p}^0, \lambda \mathbf{p}) = F(\mathbf{p}^0, p) \quad \forall \mathbf{p}^0, \mathbf{p} \in \mathbb{R}^n_{++}, \forall \lambda \in \mathbb{R}_{++}.$$

Among others, the following functions $F$ can be regarded as price indices:

$$F(\mathbf{p}^0, \mathbf{p}) = \frac{\mathbf{c} \cdot \mathbf{p}}{\mathbf{c} \cdot \mathbf{p}^0}$$

$$(\mathbf{c} = (c_1, .., c_n) \in \mathbb{R}^n_{++} \text{ constant}), \qquad (2.3)$$

$$F(\mathbf{p}^0, \mathbf{p}) = \left\{ \sum_{i=1}^n \beta_i (p_i / p_i^0)^\alpha \right\}^{1/\alpha}$$

$$(\alpha \neq 0, \beta_i > 0 \text{ real const. } \sum_{i=1}^n \beta_i = 1), \qquad (2.4)$$

$$F(\mathbf{p}^0, \mathbf{p}) = \prod_{i=1}^n (p_i / p_i^0)^{\alpha_i}$$

$$(\alpha_i > 0, \text{ real constants, } \sum_{i=1}^n \alpha_i = 1). \qquad (2.5)$$

Depending on the interpretation of the $c_i$'s, the formula (2.3) may be the price index of Laspeyres or Paasche. The functions given by (2.4) and (2.5) are well-known to be of the ACMS-type or Cobb-Douglas (CD-) type, respectively.

A characterization of the price index given by (2.3) is due to Aczél-Eichhorn [1974a], [1974b].

In this section we will show that a function $F$ satisfying the following system (2.6) of functional equations is always a price index of the CD- or of the ACMS-type.

We consider the following system for $n \geq 3$:

$$F(\mathbf{p}^0, \mathbf{p}) = \Psi^1\{ G(p_1^0, p_1), H^1( G(p_2^0, p_2), \ldots, G(p_n^0, p_n)) \}$$

$$\vdots$$

$$= \Psi^i\{ G(p_i^0, p_i), H^i( G(p_1^0, p_1), \ldots, G(p_{i-1}^0, p_{i-1}),$$

$$G(p_{i+1}^0, p_{i+1}), \ldots, G(p_n^0, p_n)) \}$$

$$\vdots$$

$$= \Psi^n\{ G(p_n^0, p_n), H^n( G(p_1^0, p_1), \tag{2.6}$$

$$\ldots, G(p_{n-1}^0, p_{n-1})) \},$$

where

$$\left.\begin{aligned}
&\Psi^i\colon \mathbf{R}_{++}^2 \to \mathbf{R}_{++}, H^i\colon \mathbf{R}_{++}^{n-1} \to \mathbf{R}_{++}, G\colon \mathbf{R}_{++}^2 \to \mathbf{R}_{++}, \\
&\Psi^i \in \mathbf{C}^1(\mathbf{R}_{++}^2), H^i \in \mathbf{C}^1(\mathbf{R}_{++}^{n-1}) \quad (i=1,2,\ldots,n). \\
&\Psi_j^i(w_1, w_2) := \partial/\partial w_j \Psi^i(w_1, w_2) > 0 \\
&(\forall w_1, w_2 > 0; \forall i = 1,\ldots,n-1; \forall j = 1,2). \\
&H_k^i(z_1,\ldots,z_{n-1}) > 0 \\
&(\forall z_1,\ldots,z_{n-1} > 0; \forall i = 1,\ldots,n; \forall k = 1,\ldots,n-1). \\
&\Psi^i(\lambda x, \lambda y) = \lambda \Psi^i(x, y) \quad (\forall x, y, \lambda > 0; \forall i = 1,\ldots,n). \\
&H^i(\lambda z_1,\ldots,\lambda z_{n-1}) = \lambda H^i(z_1,\ldots,z_{n-1}) \\
&(\forall z_1,\ldots,z_{n-1}, \lambda > 0; i = 1,\ldots,n). \\
&G(p_i^0, \lambda p_i) = \lambda G(p_i^0, p_i) \\
&\text{and } G(\lambda p_i^0, p_i) = (1/\lambda) G(p_i^0, p_i) \\
&(\forall p_i^0, p_i, \lambda > 0). \\
&G(p_i^0, p_i^0) = 1 \quad (\forall p_i^0 > 0). \\
&\Psi^i(1,1) = H^i(1,\ldots,1) = 1 \quad (\forall i = 1,\ldots,n).
\end{aligned}\right\} \tag{2.7}$$

We always denote by $f_j$ the partial derivative with respect to the $j$-th argument of a (partially differentiable) function $f\colon D \to \mathbf{R}$ ($D$ open in $\mathbf{R}^n$).

In (2.6) we require that the value of a price index remain the same no matter whether it is regarded as a function of the change of the first price and of an index of the changes of all remaining prices or whether it is a function of the change of the second price and of an index of the changes of all remaining prices,..,or whether it is a function of the change of the $n$th price and of an index of the changes of all remaining prices.

LEMMA (2.8). Every function $F: \mathbf{R}_{++}^{2n} \to \mathbf{R}_{++}$ of the form (2.6) with functions $G, \Psi^i$ and $H^i$ satisfying (2.7) is a price index.

Proof. Omitted.

LEMMA (2.9). For arbitrary, but constant $x_i = \tilde{x}_i > 0$ $(i=2,..,n)$ the following two assertions are true.

(A) The equations

$$y_1 = H^1(x_2, \tilde{x}_3,.,\tilde{x}_n)$$

and

$$y_i = H^i(x_1, \tilde{x}_2,.,\tilde{x}_{i-1}, \tilde{x}_{i+1},.,\tilde{x}_n) \quad (i=2,.,n) \tag{2.10}$$

are uniquely solvable with respect to $x_2$ and $x_1$ for

$$y_1 \in I^{y_1} \quad \text{and} \quad y_i \in I^{y_i},$$

respectively. $I^{y_k}$ $(k=1,..,n)$ are open and nonempty intervals in $\mathbf{R}_{++}$.

(B) The "inverses" $\theta^k$ $(k=1,..,n)$ of (2.10) which satisfy

$$x_2 = \theta^1(y_1)$$

and

$$x_i = \theta^i(y_i) \quad (i=2,..,n) \ (y_k \in I^{y_k}; k=1,..,n) \tag{2.11}$$

are continuously differentiable and have positive derivatives.

Proof. Omitted.

The assumptions made on $G$ in (2.7) imply

$$G(p_i^0, p_i) = \frac{p_i}{p_i^0}. \tag{2.12}$$

If we set $x_i = p_i/p_i^0$, then (2.6) becomes (for $\mathbf{x} \in \mathbf{R}_{++}^n$)

$$\Psi^1(x_1, H^1(\mathbf{x}^{\cdot 1})) = \Psi^2(x_2, H^2(\mathbf{x}^{\cdot 2}))$$

$$= \ldots = \Psi^i(x_i, H^i(\mathbf{x}^{\cdot i}))$$

$$= \ldots = \Psi^n(x_n, H^n(\mathbf{x}^{\cdot n})),$$

$$(\mathbf{x}^{\cdot i} := (x_1, .., x_{i-1}, x_{i+1}, .., x_n), \ i = 1, 2, \ldots, n). \tag{2.13}$$

This system of functional equations is an extension of the so-called "generalized associativity equation"

$$G^1(x, F^1(y, z)) = G^2(z, F^2(x, y))$$

(see Aczél [1966, p. 327]).

If we add a (technical) assumption we can prove the following.

THEOREM (2.14). Let $F$ satisfy (2.5) and (2.6). Then $F$ is locally (that is, in sufficiently small neighbourhoods of each $(\mathbf{p}^0, \mathbf{p}) \in \mathbf{R}_{++}^{2n}$) either of the form

$$F(\mathbf{p}^0, \mathbf{p}) = C \prod_{i=1}^{n} (p_i/p_i^0)^{\xi_i}$$

$$(\xi_i > 0; \ \xi_1 + \ldots + \xi_n = 1; C > 0) \tag{2.15}$$

or of the form

$$F(\mathbf{p}^0,\mathbf{p}) = \left( \sum_{i=1}^{n} b_i(p_i/p_i^0)^\alpha \right)^{\frac{1}{\alpha}} \quad (\alpha \neq 0; b_i > 0). \qquad (2.16)$$

Proof. See Gehrig, [1978, pp. 189-204].

## 3. Marketing

Distance functions defined on a set of observed characteristics of objects (individuals, goods, enterprises, etc.) play in some fields of economics an important role (marketing, taxonomy, employee selection, etc.), see Opitz [1980] or Welker [1974].

In order to arrive at a suitable classification, one needs a measure for the similarity of two objects. This measure is a function of the characteristic attributes of these objects. In the literature devoted to this subject a great variety of such distance functions were introduced (see Bock [1974] or Opitz [1980]). We mention as examples the weighted $L_r$-metric

$$d_r(\mathbf{x}^i,\mathbf{x}^j) := \left( \sum_{k=1}^{n} \gamma_k |x_{ik} - x_{jk}|^r \right)^{1/r}$$

$$r \in \mathbb{N}; \ \gamma_1,...,\gamma_n \in \mathbb{R}_+, \qquad (3.1)$$

and the Mahalanobis-function

$$d_M(\mathbf{x}^i,\mathbf{x}^j) := \sum_{k=1}^{n} \sum_{l=1}^{n} S_{k1}(x_{ik} - x_{jk})(x_{il} - x_{jl}). \qquad (3.2)$$

Most authors in marketing theory do not give a justification for the use of a special type of distance function. Therefore it is worthwhile to present characterizations of distance functions. To the best of our knowledge, the first such characterization has been given by Gehrig-Hellwig [1982] who proved the following.

THEOREM (3.3). If, and only if, a function $d: \mathbf{R}^{2n} \to \mathbf{R}$ $(n \geq 2)$ satisfies

$$d(\mathbf{x+a},\mathbf{y+a}) = d(\mathbf{x},\mathbf{y}) \quad \text{for all} \quad \mathbf{x},\mathbf{y},\mathbf{a} \in \mathbf{R}^n \qquad (3.4)$$

$$d(\lambda\mathbf{x},\lambda\mathbf{y}) = \lambda d(\mathbf{x},\mathbf{y}) \quad \text{for all} \quad \lambda > 0; \mathbf{x},\mathbf{y} \in \mathbf{R}^n \qquad (3.5)$$

$$d(\mathbf{x},\mathbf{y}) = \Psi\Big[\sum_{i=1}^{n} \chi_i(x_i,y_i)\Big], \qquad (3.6)$$

$\Psi: \mathbf{R}_+ \to \mathbf{R}$ and $\mathbf{R}^2 \to \mathbf{R}_+$ are continuous, $\Psi$ is strictly monotonic, $\Phi_i(z) := \chi_i(z,0)$ is strictly decreasing on $\{z | z \leq 0\}$ and strictly increasing on $\{z | z \geq 0\}$

$$\Phi_i(z_0) = \Phi_i(-z_0) \quad \text{for one} \quad z_0 \neq 0, \qquad (3.7)$$

then there exist constants $\eta, b_i, r \in \mathbf{R}$ with $\eta \neq 0; b_i > 0, r > 0$ such that, for all $\mathbf{x},\mathbf{y} \in \mathbf{R}^n$,

$$d(\mathbf{x},\mathbf{y}) = \eta\Big(\sum_{i=1}^{n} b_i |x_i - y_i|^r\Big)^{1/r}. \qquad (3.8)$$

## 4. Concentration measures

Aczél [1984] describes different fields of research in which functions that are closely related to the Shannon entropy

$$S^n(\mathbf{x}) = -\sum_{k=1}^{n} x_k \log x_k \quad (0 \cdot \log 0 := 0)$$

$$\mathbf{x} \in \Gamma^n := \{\mathbf{y} | y_i \geq 0 \text{ and } y_1 + .. + y_n = 1\}$$

are applied. In addition to information theory he mentions biology, ecology, psychology, detective work and, in economic theory, the measurement of inequality, where we use as a measure of concentration

$$T^n = (-\gamma)S^n \quad (\gamma > 0). \qquad (4.1)$$

Concerning the measurement of inequality Aczél argues that "in order to fight social and economic inequalities, one should explore them and measure them scientifically".

The scientific measurement means that the best way of deriving inequality measures is the axiomatic one.

In slightly modifying Kendall [1963], we can appraise the value of such a procedure as follows.

The concept of "concentration" may be very vague; the value of an axiomatic characterization is that we take the following position: we will use $T^n$ (as given in (4.1) above) as a numerical measure of "concentration" if and only if our intuitive concept of "concentration" is reasonably described by the axioms characterizing $T^n$.

From the economic point of view, every such axiom should possess a meaningful interpretation; from the mathematical point of view they must be consistent and independent.

DEFINITION (4.2). A concentration measure is a sequence of functions

$$\{K^n | K^n : \Gamma^n \to \mathbb{R}\}$$

such that

$$K^n(P\mathbf{x}) = K^n(\mathbf{x}) \tag{S}$$

for all permutation matrices $P$ and all $\mathbf{x} \in \Gamma^n$,

$$K^{n+1}(\mathbf{x};0) = K^n(\mathbf{x}) \tag{E}$$

for all $\mathbf{x} \in \Gamma^n$,

$$K^n(\mathbf{x}) > K^n(x_1,..,x_i+h,..,x_j-h,..,x_n) \tag{T}$$

for all $i,j \in \{1,..,n\} : x_j > x_i$, for all $h \in (0, \frac{1}{2}(x_j-x_i)]$ and for all $\mathbf{x} \in \Gamma^n$.

Interpretations and justifications of the axioms "symmetry" (S), "extensibility" (E) and "strict transfer principle" (T) are given in Gehrig [1984].

Note, that (S), (E) and (T) are consistent and independent (see Gehrig [1984]).

If we want to single out one special type of concentration measure we have to add further properties to (S), (E) and (T). These additional axioms refer to a special application and must also be motivated. As a first example we quote here

$$K^n(\mathbf{x}) = K^{n-1}(x_1+x_2, x_3, .., x_n)$$

$$+ (x_1+x_2)K^2(\frac{x_1}{x_1+x_2}, \frac{x_2}{x_1+x_2}) \quad (n \geq 3) \tag{F}$$

with the convention

$$0 \cdot K^2(\frac{0}{0}, \frac{0}{0}) := 0.$$

In Gehrig [1984] an interpretation of (F) and its generalization is given. Another one, from a completely different point of view, has been introduced by Buyse-Paschen [1971].

The fact that there exist at least two interpretations of (F) proves that functions satisfying this axiom together with (S), (E) and (T) are important concentration measures.

We first note that we can omit (E) in what follows.

LEMMA (4.3). If $K^n$ satisfies (S) and (F), then it also satisfies (E).

Proof. See Aczél-Daróczy [1975, p. 59].

THEOREM (4.4). A function $K^n: \Gamma^n \rightarrow \mathbf{R}$ satisfies (S), (T) and (F) if and only if

$$K^n(\mathbf{x}) = \gamma \sum_{i=1}^{n} x_i \log x_i. \tag{4.5}$$

Proof. See Gehrig [1984].

In what follows, we replace (F) by a couple of properties, namely the additivity

$$K^{2n}(x_1y,x_1(1-y),...,x_ny,x_n(1-y)) = K^n(\mathbf{x})+K^2(y,1-y),$$

$$\text{where} \quad y\in[0,1] \tag{A}$$

and the independence of irrelevant informations

$$K^n(x_1,x_2,x_3,..,x_n)-K^{n-1}(x_1+x_2,x_3,..,x_n)=H(x_1,x_2)$$

$$(n>2), \quad \text{where} \ H \text{ is defined on the set}$$

$$D^* := \{(x,y) \mid x,y>0,x+y\leq1\}. \tag{I}$$

LEMMA (4.6). The function $H$, as defined in (I), has the following properties.

(i)     $H$ satisfies (T).

(ii)    $H$ satisfies (S).

(iii)    $H$ is bounded on $D^*$.

(iv)

$$H(x,y)\begin{cases} =0 & \text{for} \ xy=0 \\ <0 & \text{otherwise.} \end{cases}$$

Proof. Omitted.

LEMMA (4.7). If $K^n$ satisfies (S), (A) and (I), and if we define

$$\Psi_y(x):=H(xy,x(1-y)) \quad (x,y \in [0,1]), \tag{4.8}$$

then, for each $x \in [0,1]$, $\Psi_y$ is a solution of the Cauchy equation

$$\Psi_y(x_1+x_2) = \Psi_y(x_1)+\Psi_y(x_2) \tag{4.9}$$

in the domain $D^*$.

Proof. See Forte-Daróczy [1968, pp. 633,634].

Now we formulate the second main result of this section.

THEOREM (4.10). A function $K^n: \Gamma^n \to \mathbb{R}$ satisfies (T), (S), (A) and (I) if and only if $K^n$ is of the form (4.5).

Proof (cf. Forte-Daróczy [1968]). By (4.6), $H$ is bounded and sign-preserving and so is $\Psi_y$. But then, according to (4.9),

$$\Psi_y(x) = x\Psi_y(1) \quad (x \in [0,1]) \tag{4.11}$$

which implies by (4.8) that

$$H(xy, x(1-y) = xH(y, 1-y) \quad (x,y \in [0,1]). \tag{4.12}$$

One derives that $K^n$ satisfies (F) and now theorem (4.4) concludes the proof, q.e.d.

## 5. Dynamic Cournot oligopoly model

As a last example we quote a simple difference equation system which has an interesting history. In 1838, the French mathematician Auguste Cournot (see Cournot [1838]) published the first exact duopoly model (duopoly means that only two firms are active on the market) and showed its global stability.

More than 120 years later Theocharis [1959] proved that Cournot obtained his result by accident, since his model becomes unstable if more than two oligopolists are on the market. It is extremely easy to derive Theocharis' result since Cournot's assumptions imply (with $\mathbf{x}(t)$ denoting the vector of expected profit maximizing outputs in the period $t$)

$$\mathbf{x}(t) = A \cdot \mathbf{x}(t-1) + b \quad z=1,2,\dots, \tag{5.1}$$

where

$$A_{(n,n)} = \begin{pmatrix} 0 & & \\ & \diagdown -1/2 & \\ -1/2 & & \\ & & \diagdown 0 \end{pmatrix} \tag{5.2}$$

As is well known, (5.1) is globally stable if and only if all eigenvalues of $A$ are smaller than 1 in absolute value. The matrix $A$ has exactly two distinct roots, namely

$$\lambda_1 = \frac{1}{2}(\text{multiplicity } n-1) \quad \text{and} \quad \lambda_2 = \frac{1-n}{2}, \tag{5.3}$$

which implies the result of Theocharis.

Although Cournot's model has attracted great attention in the literature, its complete solution has not been published; this is so much the more astonishing as this complete solution can be obtained without great difficulty.

The complete solution of (5.1) is well known to be, in the case of the existence of $(I-A)^{-1}$

$$\mathbf{x}(t) = A^t(\mathbf{x}(0)-(I-A)^{-1}\mathbf{b})+(I-A)^{-1}\mathbf{b} \tag{5.4}$$

(see Goldberg [1968, p. 327]).

As (5.4) indicates, it is necessary to determine both the $t$-th power of the matrix $A$ and the inverse of the matrix $I-A$ in order to solve (5.1) completely.

THEOREM (5.5).

$$A^t = (\tfrac{1}{2})^t \begin{pmatrix} (n-1)/n & -1/n ... -1/n \\ & \cdots & \\ & \cdots & \cdots \\ -1/n... & -1/n & (n-1)/n \end{pmatrix} + \left(\frac{1-n}{2}\right)^t \begin{pmatrix} 1/n...1/n \\ \cdots & \cdots \\ 1/n...1/n \end{pmatrix}$$

$$(I\!-\!A)^{-1} = \frac{2}{n+1} \cdot \begin{pmatrix} n & -1 & \cdots & -1 \\ -1 & \cdots & -1 & n \end{pmatrix} \tag{5.6}$$

In the first proof of (5.5) we use the fact that (5.1) is a sequence and apply an appropriate generating function method. This is the "$z$-transform" (see Jury [1964]), a method that plays in discrete systems a role analogous to the role of the "Laplace-transform" in continuous systems.

Proof. We consider the homogeneous equation

$$\mathbf{x}(t+1) = A\mathbf{x}(t). \tag{5.7}$$

Let

$$w_i(z) := \sum_{t=0}^{\infty} x_i(t) z^t \tag{5.8}$$

be the $z$-transform of $x_i(t)$. The $z$-transform of $x_i(t+1)$ is then given by

$$g_i(z) = z^{-1}(w_i(z) - x_i(0)). \tag{5.9}$$

Since we can take the $z$-transform of any vector by taking the $z$-transform of each component, (5.7) can be written in the form

$$A\mathbf{w}(z) = z^{-1}(\mathbf{w}(z) - \mathbf{x}(0)). \tag{5.10}$$

From

$$\det(I\!-\!zA) = (1-z/2)^{n-1} \cdot (1 + \frac{n-1}{2} z) \tag{5.11}$$

it follows that $(I\!-\!zA)^{-1}$ exists for all $z \neq 2, z \neq -2/(n-1)$. Taking this into account, (5.10) yields

$$\mathbf{w}(z) = (I - zA)^{-1}\mathbf{x}(0). \tag{5.12}$$

The very special form of the matrix $(I-zA)$ suggests the following approach in order to determine its inverse.

$$(I-zA)^{-1} = \begin{pmatrix} e \\ & \diagdown & f \\ & f & \diagdown \\ & & & e \end{pmatrix} \tag{5.13}$$

With this approach we indeed succeed in deriving $(I-zA)^{-1}$ and obtain (5.6) after a partial fraction expansion by inverse transformations in an elementary way.

From our Theorem the result of Theocharis, who gave the solution in the form

$$x_i(t) = A_i(\frac{1}{2}) + B_i(\frac{1-n}{2})^t + [(I-A)^{-1}\mathbf{b}]_i, \tag{5.14}$$

is easily established.

Later on, Fisher [1961] and McManus-Quandt [1961] argued that Theocharis' solution would only be correct, if the $A_i$ are interpreted not as constants but in general as polynomials of degree $n-2$ since $1/2$ is a root of multiplicity $n-1$. The Theorem shows that the $A_i$ are indeed constants which was first proved by McManus [1962], who also derived further limitations on the constants $A_i$ and $B_i$, namely

$$B_i = B \text{ for all } i=1,..,n \text{ and } \sum_{j=1}^{n} A_j = 0, \tag{5.15}$$

a result, that can easily be derived from (5.5).

In a second proof we look at (5.4) from a completely different point of view and apply "spectral decomposition". According to that,

$$A^t = \lambda_1^t E_1 + \lambda_2^t E_2, \tag{5.16}$$

where $E_1$ and $E_2$ satisfy

$$\left.\begin{array}{l} E_1+E_2 = I \\ E_1E_2 = 0. \end{array}\right\} \tag{5.17}$$

From (5.16) and (5.17) one easily derives $A^t$ and in a similar way $(I\!-\!A)^{-1}$, so that (5.5) is proved again.

Allianz Lebensversicherung -AG
Reinsburgstr. 19
D-7000 Stuttgart
West Germany

# References

[1] Aczél, J.: 1966, *Lectures on functional equations and their applications.* Academic Press, New York-London.

[2] Aczél, J.: 1984, Some recent results on information measures, a new generalization and some 'real life' interpretations of the 'old' and new measures. This volume.

[3] Aczél, J. and Z. Daróczy: 1975, *On measures of information and their characterizations.* Academic Press, New York-London.

[4] Aczél, J. and W. Eichhorn: 1974a: Systems of functional equations determining price and productivity indices. *Utilitas Math. 5,* 213-226.

[5] Aczél, J. and W. Eichhorn: 1974b: A note on additive indices. *J. Econom. Theory 8,* 525-529.

[6] Arrow, K.J., H.B. Chenery, B.S. Minhas, and R.M. Solow: 1961, Capital-labour substitution and economic efficiency. *Rev. Economics and Statistics 43,* 225-250.

[7] Bock, H.H.: 1974, *Automatische Klassifikation.* Vandenhoeck and Ruprecht, Göttingen.

[8] Buyse, R. and H. Paschen: 1971, Zur Messung der Betriebs- und Unternehmenskonzentration. *Statistische Hefte 12,* 2-13.

[9] Cobb, C.W. and P.H. Douglas: 1928, A theory of production. *Amer. Econom. Review 18,* 139-165.

[10] Cournot, A.: 1838, *Recherches sur les principes mathématiques de la théorie des richesses.* L. Hachette, Paris.

[11] Fisher, F.M.: 1961, The stability of the Cournot oligopoly solution: the effects of speeds of adjustment and increasing marginal costs. *Rev. Econom. Stud. 28,* 125-135.

[12] Forte, B. and Z. Daróczy: 1968, A characterization of Shannon's entropy. *Boll. Un. Math. Ital. 4,* 631-635.

[13] Gehrig, W.: 1976, *Neutraler technischer Fortschritt und Produktionsfunktionen mit beliebig vielen Produktionsfaktoren.* Verlag A. Hain, Meisenheim am Glan.

[14] Gehrig, W.: 1977, *Kompatibilität verschiedener Neutralitäten des technischen Fortschritts.* In *Operation Research Verfahren XXVI,* Teil 2, Verlag A Hain, Meisenheim am Glan, pp. 669-671.

[15] Gehrig, W.: 1978, Price indices and generalized associativity. In: *Theory and applications of economic indices.* Physica Verlag, Würzburg, pp. 183-205.

[16] Gehrig, W.: 1981, On the complete solution of the linear Cournot oligopoly model. *Rev. Econom. Stud.,* 667-670.

[17] Gehrig, W.: 1983, On functional equations in economic theory. *Aequationes Math. 25,* 299-306.

[18] Gehrig, W.: 1984, On a characterization of the Shannon concentration measure. This volume.

[19] Gehrig, W. and K. Hellwig: 1982, *Eine Charakterisierung der gewichteten $L_r$-Distanz.* OR-Spektrum, Springer, Berlin-Heidelberg-New York, 233-237.

[20] Goldberg, S.: 1968, *Differenzengleichungen und ihre Anwendungen in Wirtschaftswissenschaft, Psychologie und Soziologie.* Verlag R. Oldenbourg, München-Wein.

[21] Hicks, J.R.: 1932, *The theory of wages.* MacMillan Company, New York.

[22] Jury, E.I.: 1964, *Theory and applications of the z-transform method.* Wiley, New York.

[23] Kendall, D.G.: 1963, Functional equations in information theory. *Z. Wahrscheinlichkeitstheorie und verw. Gebiete 2,* 225-229.

[24] McManus, M.: 1962, Dynamic Cournot-type oligopoly models: a correction. *Rev. Econom. Stud. 29,* 337-339.

[25] McManus, M. and R.E. Quandt: 1961, Comments on the stability of the Cournot oligopoly model. *Rev. Econom. Stud. 28,* 136-139.

[26] Opitz, O. (ed.): 1978, *Numerische Taxonomie in der Marktforschung.* Franz Vahlen, München.

[27] Opitz, O.: 1980, *Numerische Taxonomie.* Gustav Fischer, Stuttgart-New York.

[28] Sato, R. and M. Beckmann: 1968, Neutral inventions and production functions. *Rev. Econom. Stud. 35,* 57-66.

[29] Shephard, R.W.: 1953, *Cost and production functions.* Princeton University Press.

[30] Theocharis, R.D.: 1959, On the stability of the Cournot solution on the oligopoly problem. *Rev. Econom. Stud. 27,* 133-134.

[31] Uzawa, H.: 1961, Neutral inventions and the stability of growth equilibrium. *Rev. Econom. Stud. 28,* 117-123.

[32] Welker, R.B.: 1974, Discriminant analysis as an aid to employee selection. *Acc. Rew.,* 514-523.

L. Reich

# On multiple uses of some classical equations

In Complex and Real Analysis many classes of important functions can be defined as solutions of functional equations which satisfy certain regularity conditions (e.g. analyticity or holomorphy in the complex case, convexity in the real case). A striking example was the restructuring of the theory of the gamma function by E. Artin which made the theory in the real case crystal clear and esthetically pleasing and served as a pattern for developing similar theories for other special functions. In complex analysis we may consider the theory of automorphic and modular functions as a chapter which is governed by systems of functional equations, and this is even more evident for the zeta functions (in number theory and algebraic geometry). In the investigation of these functions their functional equations play a decisive role since all interesting analytic properties of these functions are derived by studying the equations. Just recently, the dilogarithm became interesting in algebraic topology and in topology of 3-dimensional manifolds. Many of the functional equations satisfied by the dilogarithm are now of use for this purpose, after having been forgotten for years. But also the theory of elliptic functions (including the trigonometric functions as degenerate cases) and that of theta functions could be mentioned as proof for our statement.

During the last years iteration problems have become increasingly interesting. Among them, the determination of iterative roots and embedding problems of mappings into one parameter groups (flows) are closely related to classical functional equations: to Babbage's equation and to the famous translation equation, which is an equation in several variables. The task of solving the embedding problems explicitly leads to other famous classical equations which are equations in one variable, namely to the Abel equation, to Schröder type equations, but also to some functional-differential

53

J. Aczél (ed.), Functional Equations: History, Applications and Theory, 53-54.
© 1984 by D. Reidel Publishing Company.

equations and to difference systems. On the other hand, the translation equation is used in the study of algebraic and geometric objects, of automata, and in dynamics.

There is no lack of applications of functional equations as a tool in the axiomatization of various theories. Information theory has already been mentioned by previous authors as an outstanding example. In the mathematical foundations of computer science there occur difficult and interesting functional equations for convergent as well as formal power series.

Here we may return to the role of equations in iteration theory. Also here, as well as in some other fields, it is often useful to look for solutions in the form of formal power series. This in turn proved to be a powerful incentive for further developments of the theory of such series. One example of this direction of research is the theory of formal groups, although it is usually not considered from the point of view of functional equations.

Mathematisches Institut
Universität Graz
Brandhofg. 18
A-8010 Graz
Austria

A. Sklar

## On some unexpected theoretical and practical applications of functional equations

Since functional equations arise in virtually every area of mathematics, both pure and applied, it is not surprising that the study of such equations can lead to useful results. For example, in the cosmological theory of Milne (now slowly coming back into favour as a subject of investigation after some years of neglect) it is necessary to correlate signals exchanged between different "observers". To effect such correlation, one must be able to find iterative square roots of certain functions, where an iterative square root of a function $g$ is a function $f$ such that $f(f(x))=g(x)$ for all $x$ in the domain of $g$.

But even the study of the theory of functional equations for its own sake can have far-reaching consequences. The outstanding example, at least in pure mathematics, is the notion of a Hamel basis, which was introduced precisely in order to characterize the general solution of the Cauchy functional equation. In number theory, the beautiful theory of Hecke involves a continual interplay between functional equations, their sources, and their consequences.

Perhaps the most useful service that the theory of functional equations can perform for both mathematicians and users of mathematics is to prevent their being unintentionally misled. For example, it is customary in preparing single differential equations or systems for numerical treatment to replace them by difference equations. This has led many people to look at difference equations, interpreted as discrete dynamical systems, with a view towards gaining insight into continuous dynamical systems and the differential equations governing them. Thus many people, including meterologists and ecologists, have considered the discrete dynamical system given by the equation

$$x_{n+1} = 2\lambda x_n(1-x_n), \quad n=0,1,2,...$$

55

J. Aczél (ed.), Functional Equations: History, Applications and Theory, 55-56.
© 1984 by D. Reidel Publishing Company.

with $0 < \lambda \leq 2$ and the $x$'s in the unit interval $[0,1]$. Yet the theory of functional equations shows that, if $\lambda > 3/2$ (which includes the cases of interest), then this discrete dynamical system cannot be embedded in any continuous dynamical system whatever. Thus this example is of no relevance for the study of continuous dynamical systems, which is what most of the people who have worked with the example are really interested in. And this example is typical, not exceptional.

I end with an anecdote. Some months ago, a friend of mine came to me with a problem that arose in his consulting work in the Route 128 area around Boston. It involved the transduction of information from one sensory mode to another, and led to a messy-looking equation. I was able to show that the equation could be reduced to the well-known Schröder functional equation. I directed my friend's attention to some works dealing with this equation, in particular the book "Functional Equations in a Single Variable" by M. Kuczma. My friend, finding that the book was out-of-print and difficult to obtain, borrowed my copy, and later reported significant progress.

Department of Mathematics
Illinois Institute of Technology
Chicago, IL 60616
U.S.A.

A. Krapež

## Functional equations on groupoids and related structures

Here we describe a uniform method for solving *any* system of functional equations. The method is a joint work of S.B. Prešić and the author and is a generalization of methods used in [1], [2], [3], and [4].

In [1] and [2] functional equations of generalized associativity, bisymmetry, transitivity and distributivity are solved. Methods used are generalized further in [3]. Independently of [3], S.B. Prešić in [4] developed an even more general method.

The core of all these methods is the construction of an equivalence which should be the kernel of both the left hand and the right hand side of the functional equation. Here we denote this equivalence as $\sim$; $\sim$ is the most important of the triple $(\sigma, \sim, *)$, called a resolvent. Here is its definition.

A triple $(\sigma, \sim, *)$ (where $\sigma$ is a unary relation and $\sim$ a binary relation while $*$ is a unary operation) is called a resolvent if the following conditions are satisfied ($\emptyset \neq S \subset T$):

(1)  $\sigma a$ for all $a \in S$,

(2)  $\rceil \sigma a$ for all $a \in T \setminus S$,

(3)  $a \neq b$ for all different $a, b \in S$,

(4)  $\sim$ is an equivalence on $T$,

(5)  $\sigma x, \sigma y, x \sim y \rightarrow x = y$,

(6)  $x \sim x^*$,

(7)  $\sigma x^*$.

It is easy to see that $\sigma$ separates elements of a subset $S$ of a set $T$, that the restriction of the $\sim$ to $S$ is equality on $S$ and that $*$ is an operation of $T$ such that $x^*$ is the unique element of the $\sim$-class of $x$ which belongs to $S$.

J. Aczél (ed.), Functional Equations: History, Applications and Theory, 57-64.
© 1984 by D. Reidel Publishing Company.

Conditions (1)-(7) with a reformulation of a system of equations form a theory equivalent to the given system of equations. Here is how we transform the given system of equations.

Let a system $\Gamma$ of functional equations be given. All functions of $\Gamma$ are either given (we collect them in the set $\Pi$ and call them parameters) or unknown. Of all unknown functions some are main operations in at least one term of some equation from $\Gamma$ and they form the set $\Omega_0$. All the other unknown operations belong to the set $\Omega_1$. $\Omega_0 \cup \Omega_1 = \Omega$, the set of all unknown functions from $\Gamma$.

For example, for the functional equation

$$A(x+y,z) = E(F(x+z,u),A(y+u,x)) \tag{8}$$

on the set $\mathbb{R}$ of reals, $\Pi = \{+\}$, $\Omega_0 = \{A,E\}$ and $\Omega_1 = \{F\}$. $A$ and $E$ are main operations of $A(x+y,z)$ and of $E(F(x+z,u),A(y+u,x))$, repectively.

The first step in transforming a system $\Gamma$ is the substitution of new variables for every subterm of $\Gamma$. So for every equation of $\Gamma$ we have an equivalent implication.

For equation (8), the corresponding implication is

$$x+y = v, \quad x+z = w, \quad F(w,u) = p, \quad y+u = q,$$
$$A(q,x) = r \Longrightarrow A(v,z) = E(p,r). \tag{9}$$

In such a manner we obtain the system $\Gamma_=$. By $\Gamma_\sim$ we denote the system obtained from $\Gamma_=$ by substituting $(x_1,...,x_n,F)\sim v$ for $F(x_1,...,x_n)=v$ for all $F \in \Omega_0$. To all antecedents of implications we add formulas $\sigma v$ for all variables $v$ from that implication.

For (8), implication (9) is transformed into:

$$\sigma x, \ \sigma y, \ \sigma z, \ \sigma u, \ \sigma v, \ \sigma w, \ \sigma p, \ \sigma q, \ \sigma r, \ x+y = v,$$
$$x+z = w, \quad F(w,u) = p, \quad y+u = q,$$
$$(q,x,A)\sim r \Longrightarrow (v,z,A)\sim(p,r,E).$$

Let also

$$S_F = \{(x_1,...,x_n,F)|x_1,...,x_n \in S; \ F \in \Omega_0, \ F \ n\text{-}ary\}$$

and

$$T = S \cup ( \bigcup_{F \in \Omega_0} S_F).$$

We say that $(\sigma, \sim, *)$ is satisfying (1)-(7) and $\Gamma_{\sim}$ is a $\Gamma$-resolvent.

THEOREM 1. A system $\Gamma$ of functional equations has a solution on $S$ iff the theory (1)-(7), $\Gamma_{\sim}$ is consistent.

Proof. (i) If the system $\Gamma$ has a solution, then the above constructions are carried out in some model of (1) - (7), $\Gamma_{\sim}$ and consequently this theory is consistent.

(ii) Let $M$ be a model of consistent theory (1) - (7), $\Gamma_{\sim}$ in the language $L = \Pi \cup \Omega_1 \cup \{\sigma, \sim, *\}$.

For $F \in \Omega$ we define

$$F(x_1, ..., x_n) = \varsigma^{-1} F(\varsigma x_1, ..., \varsigma x_n)$$

$$\text{for } F \in \Omega_1,$$

$$F(x_1, ..., x_n) = \varsigma^{-1} ((\varsigma x_1, ..., \varsigma x_n, F)^*)$$

$$\text{for } F \in \Omega_0,$$

where $\varsigma$ is an interpretation of elements of $T$ in $M$. Condition (3) ensures that $\varsigma|S$ is an injection and the above definitions of operations from $\Omega$ force $\varsigma$ to be an isomorphism of $(S, \Pi, \Omega)$ and $(\varsigma(S), \Pi, \Omega)$. We shall identify these structures.

If $F(t_1, ..., t_m)$ is a subterm of $\Gamma$ then, using antecedents of implications from $\Gamma_{\sim}$, we can replace it by a variable from $\Gamma$.

Let us assume that all the subterms $t_1, ..., t_m$ of $F(t_1, ..., t_m)$ are replaced by variables $v_1, ..., v_m$, respectively. If $F \in \Pi$ or $F \in \Omega_1$ then there is a variable $v$ such that the formula $F(v_1, ..., v_m) = v$ is among the antecedents of some implication in $\Gamma_{\sim}$, so we can replace $F(t_1, ..., t_m)$, i.e. $F(v_1, ..., v_m)$ by $v$. If $F \in \Omega_0$, then there is a variable $v$ such that the formula $(v_1, ..., v_m, F) \sim v$ is among the antecedents of some application in $\Gamma_{\sim}$, so, using (b) and $\sigma v$, we can replace

$F(t_1,...,t_m)$, i.e. $F(v_1,...,v_m)$ by $v$. Finally, the consequence of the implication we were using ensures that the left hand side and right hand side of a particular equation from $\Gamma$ are equal.

Since all the other equations are satisfied in a similar way, $M$ indeed defines a solution.

THEOREM 2. If the system $\Gamma$ of functional equations has at least one solution on a set $S$, then the general solution of $\Gamma$ on $S$ is given as follows:

(a)  $F$ is arbitrary for $F \in \Omega_1$ ,

(b)  $F(x_1,...,x_n,F)^* = (x_1,...,x_n,F)^*$ for $F \in \Omega_0$.

Proof. (i) If we define functions from $\Omega$ by (a) and (b), then the proof that this defines a solution of $\Gamma$ is essentially the same as (ii) of Theorem 1.

(ii) Let a solution of $\Gamma$ on $S$ be given. Let $T = S \cup (\bigcup_{F \in \Omega_0} S_F)$ and let $\sigma, *$ and $\sim$ be defined on $T$ in the following way:

$$\sigma x \text{ iff } x \in S,$$

$$x^* = x \text{ for } x \in S,$$

$$x^* = F(x_1,...,x_n) \text{ for } x = (x_1,...,x_n,F) \in S_F,$$

$$x \sim y \text{ iff } x^* = y^*.$$

Then (1)-(7), $\Gamma_\sim$ and (a), (b) are all satisfied so we indeed have a general solution of $\Gamma$.

It is easy to see that the following corollary is also valid.

COROLLARY. If a system $\Gamma$ has a solution on $S$ then, for given $F \in \Omega_1$ and any $\Gamma$-resolvent $(\sigma, \sim, *)$, the solution given by (a) and (b) is unique.

EXAMPLE 1.  Let the equation

$$A(B(x,y),A(x,z)) = A(y,B(y,z)) \tag{10}$$

be given and let $S=\{0,1\}$.  Then

$$T = \{0,1,(0,0,A),(0,1,A),(1,0,A),(1,1,A)\}$$

and $\Gamma_\sim$ is:

$$\sigma x,\ \sigma y,\ \sigma z,\ \sigma u,\ \sigma v,\ \sigma w,\ \ B(x,y) = u,\ (x,z,A)\sim v,$$

$$B(y,z) = w \Longrightarrow (u,v,A)\sim(y,w,A).$$

We will find solutions of (10) in the case when $B$ is the implication.

Substituting 0 and 1 for $x$ and $y$, we obtain:

$$\sigma z,\ \sigma v,\ (0,z,A)\sim v \Longrightarrow (0,1,A)\sim(1,v,A)\sim(1,z,A)$$

$$\sigma z,\ \sigma v,\ (1,z,A)\sim v \Longrightarrow (0,0,A)\sim(0,1,A)$$

$$\sigma z,\ \sigma v,\ (1,z,A)\sim v \Longrightarrow (1,v,A)\sim v.$$

If $(1,1,A)\sim0$ $((1,1,A)\sim1$ and $(0,0,A)\sim1)$ we obtain unique trivial solution $A(x,y)\equiv0$ $(A(x,y)\equiv1)$.

If $(1,1,A)\sim1$ and $(0,0,A)\sim0$ then $(0,1,A)\sim1$ and $(1,0,A)\sim1$ and

$$* = \begin{pmatrix} 0 & 1 & (0,0,A) & (0,1,A) & (1,0,A) & 1,1,A) \\ 0 & 1 & 0 & 1 & 1 & 1 \end{pmatrix}$$

defines the solution $A(x,y)=x\vee y$.

EXAMPLE 2.  Let $\{0,1\}\subset S$ and $A(x,y)\equiv0$, $E(x,y)\equiv1$.  The equation

$$A(x,y) = E(B(x,y),y)$$

has no solution.  We can see this also from $\Gamma_\sim$:

$$\sigma x,\ \sigma y,\ \sigma z,\ (x,y,B){\sim} z \Longrightarrow A(x,y) = E(z,y),$$

i.e.,

$$\sigma x,\ \sigma y,\ \sigma z,\ (x,y,B){\sim} z \Longrightarrow 0 = 1.$$

If we take $z=(x,y,B)^*$, we will get a contradiction. By Theorem 1 $\Gamma$ has no solution.

EXAMPLE 3. Let $\Gamma$ be the Cauchy equation:

$$f(x+y) = f(x)+f(y)$$

on the set $N$ of natural numbers. Then $\Gamma_{\sim}$ is

$$\sigma x, \sigma y, \sigma z, \sigma u, \sigma v, x+y = z,$$

$$(x,f){\sim} u,\ (y,f){\sim} v \Longrightarrow (z,f){\sim} u+v.$$

For $x=0$, $y=0$, $z=0$, $u=(0,f)^*$ and $v=(0,f)^*$ we get $(0,f){\sim}(0,f)^*+(0,f)^*$. Since $(0,f)^*+(0,f)^*$ is a natural number, that is, $\sigma((0,f)^*+(0,f)^*)$, we have $(0,f)^*{\sim}(0,f)^*+(0,f)^*$ and $(0,f)^*=(0,f)^*+(0,f)^*$. This is possible for $(0,f)^*=0$ only. so $(0,f){\sim}0$.

Let $\mathbf{a}=(1,f)^*$ (so $\mathbf{a}$ is a natural number). We shall prove $(n,f){\sim} n\mathbf{a}$ for all natural numbers $n$. For $n=1$ this is true.

Suppose that $(n,f){\sim} n\mathbf{a}$ for some natural $n$. For $x=n$, $y=1$, $z=n+1$, $u=n\mathbf{a}$, $v=\mathbf{a}$ from $\Gamma_{\sim}$ it follows:

$$(n,f){\sim} n\mathbf{a},\ (1,f){\sim}\mathbf{a} \Longrightarrow (n+1,f){\sim} n\mathbf{a}+\mathbf{a}.$$

Using the induction hypothesis we get $(n+1,f){\sim}(n+1)\mathbf{a}$. So $(n,f){\sim} n\mathbf{a}$ for all natural numbers $n$.

Since $n\mathbf{a}$ is a natural number i.e. $\sigma(n\mathbf{a})$, it follows that $(n,f)^*=n\mathbf{a}$ and the general solution of the Cauchy equation on $N$ is $f(x)=\mathbf{a}x$ for $\mathbf{a}\in N$.

EXAMPLE 4. Let $\Gamma$ be

$$A(B(x,y),B(x,z)) = E(y,B(x,u)).$$

Then $\Gamma_{\sim}$ is

$$\sigma x,\ \sigma y,\ \sigma z,\ \sigma u,\ \sigma p,\ \sigma q,\ \sigma r,\ B(x,y) = p,$$
$$B(x,z) = q,\ \ B(x,u) = r \implies (p,q,A) \sim (y,r,E).$$

If $S = \{0,1\}$ and $B$ is implication, we have

$$p = q = r = 1$$

and

$$(1,1,A) \sim (y,1,E)\ \ \text{for}\ \ x{=}0,$$
$$p = y,\ \ q = z,\ \ r = u$$

and

$$(y,z,A) \sim (y,u,E)\ \ \text{for}\ \ x{=}1.$$

It follows that all members of $T \backslash S$ are in the same class of $\sim$. There are exactly two solutions:

$$A(x,y) = E(x,y) \equiv 0,$$
$$A(x,y) = E(x,y) \equiv 1.$$

The situation is similar in the general case. If any operation from $\Omega_0$ appears in $\Gamma$ as a main operation only and all main operations are in $\Omega_0$, then the general solution of $\Gamma$ is given by the equivalence $\sim$ on $T \backslash S$, satisfying $\Gamma_{\sim}$ and by the function $f\colon (T \backslash S)_{/\sim} \to S$ (S.B. Prešić [4]).

Theorem 1 and Theorem 2 can be generalized further if we replace equations from $\Gamma$ by inequalities or some other relations. Also, functions from $\Omega$ need not be groupoids.

Matematički Institut
Knez Mihailova 35
Y-11000 Beograd, Yugoslavia

## References

[1]   Krapež, A.: 1980, Generalized associativity on groupoids. *Publ. Inst. Math. Beograd 28 (42)*, 105-112.

[2]   Krapež, A.: 1981, Functional equations of generalized associativity, bisymmetry, transitivity and distributivity. *Publ. Inst. Math., Beograd 30 (44)*, 81-87.

[3]   Krapež, A.: 1982, Generalized balanced functional equations on $n$-ary groupoids. *Proceedings of the symposium n-ary structures*, Skopje, 13-16.

[4]   Prešić, S.B.: 1982, A general solving method for one class of functional equations. *Proceedings of the symposium n-ary structures*, Skopje, 1-9.

# Papers

J.G. Dhombres

# Some recent applications of functional equations

It is difficult to provide an informative survey of all the works which recently appeared in the theory of functional equations and it is obviously impossible to comment on the various applications in other mathematical domains or outside mathematics. In this respect, the best thing is to look at the recent proceedings of Symposia in functional equations (see [1], [2] and [26]).

Therefore I would like to select some *particular* topics and present results, some of which are not yet published. My subjective selection has been made with the theme of the present conference in mind and thus with an emphasis on topics close to or generated by *applications,* clearly under the limitations of my own awareness of new results. However, we must never forget that curiosity for its own sake always piques the mathematician and that from the most practical problem he seems often able to derive an abstract point of view. I am not sure that in doing so he totally deserts the service of man. In this respect, my selection of topics has also been made so to include open problems. In order to help the reader and because applications of various origins are not easy to organize in a logical way, I have named headings according to purely mathematical items. We have four different kinds of applications in mind, chosen to provide examples of four different roles of functional equations. With the first heading, *functional equations for set functions,* we wish to deal with *chronogeometry,* a subfield of relativity theory. In this case, the help brought by functional equations is typical of the clarification of a possible axiomatic set-up. The second heading, *means and functional equations,* explains the role of functional equations as a classification tool for studying various kinds of means in various settings. Some means have interesting applications to averaging theory for the partial differential equations governing turbulence in water pipes. Some other means lead

J. Aczél (ed.), Functional Equations: History, Applications and Theory, 67-91.
© 1984 by D. Reidel Publishing Company.

to geometrical problems. The third heading, *functional equations and group theory*, shows how functional equations can be used for finding subgroups of groups of operators. The fourth and last heading deals with *conditional functional equations*, a fast developing field in functional equations, which shows the role of some limitations of validity of the equation by restricting the domains of variables. We have selected an application from ideal gas theory and we have explained some consequences for the geometry of Banach spaces.

## 1. Functional equations for set functions

Functional equations were classically defined and solved for variables belonging to the real line or to more complex structures. Recently functional equations have occurred in an interplay between subsets of a set and elements of the set. This interplay appears to be quite rich and complex. It is interesting to note that the equations originate in relativity theory (see [3] and [4]). The idea is to define on an affine $n$-space a geometry based on a relation of precedence of points. We associate with one point all those upon which this point can act: a cause-effect structure. For such a study the word *chronogeometry* was coined.

Every event $x$ is the vertex of a cone $C_x$ of events preceded by $x$. Relevant to relativity theory is the hypothesis that $C_x$ is just the translate of a given cone $C$ with vertex at the origin. Thus $C_x = C + x$ (the surface of $C_x$ is the "light cone at $x$"). For the *precedence relation* to be transitive, it is enough to suppose the cone $C$ to be convex. Some metric properties (spatial and temporal structures) can be deduced from the cones $C_x$.

In this context, the interesting mappings $f: \mathbf{R}^n \to \mathbf{R}^n$ are those which keep the cones invariant, i.e. $f(C_x) = C_{f(x)}$. A functional equation characterizes this property:

$$C + f(x) = f(C + x) \quad (x \in \mathbf{R}^n). \tag{1}$$

By purely geometrical arguments (and with no assumption on the continuity of $f$), the following can be proved (see Alexandrov [3]).

THEOREM 1. Let $n>2$ and suppose that $C$ does not lie in a plane, that its closed convex hull contains no straight line and that this same closed convex hull is not the cartesian product of a ray and an $(n-1)$-dimensional cone. Then a bijection $f: \mathbf{R}^n \to \mathbf{R}^n$ with $f(0)=0$ which is a solution of (1), is linear.

A corollary asserts that Lorentz transformations are the bijections preserving the system of elliptic cones in $\mathbf{R}^4$, showing that such transformations are fully determined by the light cones.

As usual we may investigate a Pexider analogue of equation (1), namely

$$C+g(x) = f(C+x). \tag{2}$$

Under more restrictive conditions for $C$, a bijection $f$ with $f(0)=0$ which is a solution of (2) has to be linear.

For $\mathbf{R}^n$, or for more general structures, the general solution of (1) has not been investigated when the cone $C$ is replaced by a given nonempty set $E$. It would be nice to get at least necessary and/or sufficient conditions for $E$ in order that the general solution of (1) should be affine.

It is obvious that adding a constant to $f$ preserves the fact that $f$ is a solution of (1). We may therefore suppose $f(0)=0$, in which case $f(C)=C$. Equation (1) leads then to the following functional equation

$$f(x+C) = f(x)+f(C). \tag{3}$$

We will now deal with this equation under the hypothesis that $C$ is a convex cone. In this case, $x+C$ is a subset of $C$ when $x$ belongs to $C$. We may suppose $f$ to be defined only on $C$.

THEOREM 2. Let $C=\mathbf{R}_+=]0,\infty[$. Suppose $f: C \to \mathbf{R}$ is a nonidentically zero function, the image of which is connected but different from $\mathbf{R}$. Then $f$ is a solution of (3) if and only if $f$ is a homeomorphism from either $C$ onto $C$ or $C$ onto $-C$.

As a consequence we easily get the set of all continuous solutions of (3); it consists of four different types:

$f=0$,

$f$ is strictly increasing and continuous from $C$ onto $C$,

$f$ is strictly decreasing and continuous from $C$ onto $-C$,

$f$ is continuous and $f(x+C)=\mathbf{R}$ for all $x$ in $C$.

This Theorem 2 can be adapted to get a characterization of monotonic functions on an interval of $\mathbf{R}$ (for details, see Chapter 9 of [5]). It seems possible to get a generalization of Theorem 2 when $C$ is a nonempty open convex cone of $\mathbf{R}^n$ for $n>1$. Some regularity properties were obtained by Forti in [6] and [7]. At least, we may obtain the set of all continuous solutions:

THEOREM 3. A continuous $f: C \rightarrow \mathbf{R}$, where $C$ is an open, convex and nonempty cone of $\mathbf{R}^n$ with vertex at 0, satisfies (3) for all $x$ in $C$ if and only if $f$ is of one of the following four mutually exclusive forms: $f(C)=\{0\}$; $f(C)=]0,\infty[$ and $f$ is strictly increasing for the order generated by $C$, unbounded from above on any ray of $C$; $f(C)=]0,-\infty[$, $f$ is strictly decreasing for the order generated by $C$ and unbounded from below on any ray of $C$; $f(C+x)=\mathbf{R}$ for all $x$ in $C$.

To extend the results of Theorems 2 and 3 to the case where $f$ takes its values in $\mathbf{R}^m$ $(m>1)$ appears to be rather cumbersome. It is clear that, if

$$f = (f_1,\ldots,f_m)$$

is a solution of (3) for $f\colon C\to \mathbf{R}^n$, then $f_i$, for all integers $i=1,2,\ldots,m$, is a solution of (3) for $f_i\colon C\to \mathbf{R}$. But the converse is not true and more has to be said on the relations between the functions $f_i$. See [8] for the following result and for more.

THEOREM 4. Let $f_i\colon C\to \mathbf{R}$ $(i=1,\ldots,m)$ be continuous solutions of (3), satisfying the hypothesis of Theorem 2 where $C$ is an open convex and nonempty cone of $\mathbf{R}^2$, with vertex at 0. Then $f=(f_1,f_2,\ldots,f_m)$ is a solution of (3) with $f\colon C\to \mathbf{R}^m$ if and only if, for all $(x,y)$ in $C$, $f_i(x,y)=a_i g(x)$ (or for all $(x,y)$ in $C$, $f_i(x,y)=a_i g(y)$) where $a_i\neq 0$ and where $g$ is a continuous solution of (3) satisfying the hypothesis of Theorem 2.

Almost nothing is known about the set $S(C)$ of all (possibly regular) solutions of equation (3) when $C$ is not a cone. An interesting question would be to compare two subsets $C_1$ and $C_2$ for which

$$S(C_1) = S(C_2).$$

Along this line of investigation, it is possible, however, to construct an example where continuity is the required regularity for $f$ and where the two homomorphic subsets $C_1$ and $C_2$ are such that $S(C_1)$ differs from $S(C_2)$.

Perhaps it is then better to replace equation (3) by a family of functional equations? For example, we can study

$$f(\lambda D+x) = f(\lambda D)+f(x) \text{ for all } \lambda>0, x>0, \tag{4}$$

where $D$ is a given nonempty subset of $]0,\infty[$. In fact, a solution of (4) is a particular solution of a functional equation of type (3), where the cone $C$ occurring in (3) is the cone generated by the set $D$, i.e. here $C=]0,\infty[$. The family $S(D)$ of all solutions of equation (4), having the properties stated as hypotheses in Theorem 2, is independent from the nonempty set $D$. In fact $S(D)$ is exactly the family of the linear functions $f(x)=\alpha x$ where $\alpha\geq 0$. However, if less regularity with regard to the solution $f$ is required, then the convenient tool which allows us to distinguish between two sets $D$ solutions of equation (4) is the *degree of rarefaction* of $D$ (as shown in [3]). This degree $d(D)$ is defined as the upper bound of the ratios $b/a$ where $a$ and $b$ are positive, $a \leq b$ and $]a,b[$ is a subset $]0,\infty[\backslash D$. If $D$ is bounded from above, then $d(D) = \infty$. We shall only consider the family **D** of all subsets $D$ of $]0,\infty[$ for which the degree $d(D)$ is attained for no ratio $b/a$.

THEOREM 5. For two subsets $D_1$ and $D_2$ in **D**, the set of all solutions of equation (4) bounded from below, coincide if and only if $d(D_1) = d(D_2)$.

Equation (3) is related to the Cauchy functional equation. In the same way we may get new functional equations containing a given set from other classical functinal equations. Some results were obtained for the Jensen equation (see [9] and [10]). Both for the Jensen equation, the Cauchy exponential functional equation and for the d'Alembert equation, results have recently been found but we have no room here to give the exact results (cf. [11]).

## 2. Means and functional equations

Among the oldest tools both in mathematics and in domains related to mathematics are the *means* of two (or more) numbers: arithmetic, geometric or harmonic means for example. Such means are special cases of the so-called regular *quasi-arithmetic mean* $M(x,y)$:

$$M \cdot (x,y) = f^{-1}\left(\frac{f(x) + f(y)}{2}\right), \tag{5}$$

where $f$ is some strictly montonic and continuous function on an interval. An axiomatic characterization of regular quasi-arithmetic

means on intervals was simultaneously provided by A. Kolmogorov
and N. Nagumo in 1930. We quote here the nice result obtained by
J. Aczél (for a proof see [12] or [5]). A binary law defined on a set $I$
means a mapping $M:\ I\times I\to I$.

Let $I$ be an interval of $\mathbf{R}$. A binary law defined on $I$ is a
regular quasi-arithmetic mean if and only if the following conditions
are satisfied:

$$M(x,x) = x \text{ for all } x \text{ in } I, \tag{6}$$

$$M(x,y) = M(y,x) \text{ for all } x,y \text{ in } I, \tag{7}$$

$$M(M(x,y),M(z,w)) = M(M(x,z),M(y,w)) \tag{8}$$

$$\text{for all } x,y,z,w \text{ in } I,$$

$$M \text{ is continuous}, \tag{9}$$

$$M \text{ is strictly increasing in each of its arguments.} \tag{10}$$

There have been many successful attempts to generalize this
result. An easy job is to have more than two variables $x$ and $y$.
Another rather easy job is to avoid the symmetry condition (7) and
to use weighted quasi-arithmetic means. More difficult is to
consider variables $x,y$ belonging to $\mathbf{R}^n$ because then we have to
replace the monotonicity condition (10) by some other condition.
K. Sigmon ([13]) proved that with $I$ an open or closed $n$-cell $I\subseteq \mathbf{R}^n$,
if we replace condition (10) by the cancellative property for the
binary law $M$, mutatis mutandis the same result remains valid in
the sense that in (5) $f$ is now a homeomorphism over $I$. Only the
case of an $n$-cell of $\mathbf{R}^n$ has been investigated. It could be interesting
to look for some other subsets, even on $\mathbf{R}$ (a Cantor set for
example).

The representation (5) for $M$ is almost unique. More precisely,
for two functions $f$ and $g$, for which

$$M(x,y) = f^{-1}\left(\frac{f(x)+f(y)}{2}\right) = g^{-1}\left(\frac{g(x)+g(y)}{2}\right), \tag{11}$$

there exist constants $a \neq 0$ and $b$ and

$$f(x) = ag(x)+b.$$

In other words, $g$ is determined when values of $g$ are given at two distinct given points of $I$ (Sturm-Liouville conditions). One may write (11) in a different way as

$$g(M(x,y)) = A(g(x),g(y)), \qquad (12)$$

where $M$ is a quasi-arithmetic mean as defined by (5) and $A$ the usual arithmetic mean. Uniqueness results under Sturm-Liouville conditions have been given for functional equations similar to or slightly more general than (12) but under the restriction that $g$ be continuous (see [14] and [15]). Very little has been added in recent years. We may undertake the study of (12) in a more general setting. Let $I$ be a nonempty set. If we find a convex subset $J$ of a real linear space $E$ and a bijection $f\colon I \to J$, we may define a *quasi-arithmetic mean* (q.a. mean) $M(x,y)$ on $I$ by using (5). Let $B(I)$ be the set of all bijections $f\colon I \to J$ for all possible $E$ and $J$ with the properties as described. For $f_1\colon I \to J_1$ and $f_2\colon I \to J_2$, where $f_1$ and $f_2$ are in $B(I)$, we define $f_1 \sim f_2$ if there exists a Jensen function $h\colon J_1 \to J_2$ for which $f_2 = h \circ f_1$. Recall that by definition a function $h$ is *Jensen* if, for all $x, y$ in $J_1$,

$$h\left(\frac{x+y}{2}\right) = \frac{h(x)+h(y)}{2}.$$

This relation $\sim$ is an *equivalence relation* on $B(I)$. Moreover one can prove that the functions $f_1$ and $f_2$ in $B(I)$ define the same q.a. mean on $I$ if and only if $f_1 \sim f_2$.

In some cases, the relation $\sim$ can be made more precise. Let $E_f$ be the linear subspace generated by $J-b$ in $E$, where $b \in J = f(I)$. We say that $f_2 \in B(I)$ is an $f_1$ -*regular function*, for $f_1 \in B(I)$, if there exist a linear and onto operator $[A]\colon E_{f_1} \to E_{f_2}$, and vectors $a$ in $f_1(I)$, $b$ in $f_2(I)$ such that

$$f_2(x) = [A](f_1(x)-a) + b \qquad (x \in I).$$

Let $Q(I)$ be the subset of all $f$ in $B(I)$ for which $E_f$ is finite-dimensional and $f(I)=J$ is included in a *quadrant*. Recall that a $(n-)quadrant$ of a finite $(n-)$dimensional space $E$ is the set of all $x$ in $E$ such that, for some $a$ in $E$, $x_i >_i a_i$, $i=1,2,...,n$, where $>_i$ means $>$ or $<$, and where $x_i$ (resp. $a_i$) are the coordinates of $x$ (resp. $a$) in some basis of $E$. In $E$ we also define an *open band* as the set of all $x$ in $E$ such that for some basis and some $i$, $i=1,...,n$, we have $\beta >_i x_i >_i \alpha$, where $\alpha$ and $\beta$ are real numbers or possibly $\pm\infty$. If only one value is infinite, we get an *half-plane*.

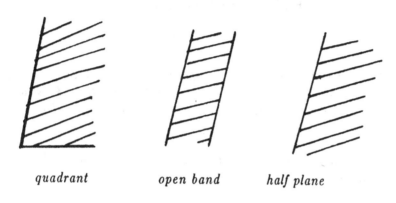

quadrant          open band          half plane

PROPOSITION 6. Let $f_1 \in Q(I)$ and $f_2 \in B(I)$. Then $f_2 \sim f_1$ if and only if $E_{f_1}$ and $E_{f_2}$ have the same (finite) dimension and $f_2$ is an $f_1$-regular function.

We solve (12) in this general context but with $M=A$, a q.a. mean on $I$. We then call $g: I \to I$ an *M-affine function* if, for all $x,y$ in $I$,

$$g(M(x,y)) = M(g(x),g(y)). \qquad (13)$$

We say that $g: I \to I$ is an *f-affine function* if there exists a linear operator $[A]: E \to E$ and a vector $a \in E$ such that $f(g(x))=[A]f(x)+a$ for all $x \in I$.

THEOREM 7. Let $I$ be a nonempty open convex subset of a real linear space of dimension $n$, with $n>1$. Let $f: I{\rightarrow}I$ be a homeomorphism defining a q.a. mean $M$ on $I$. A necessary and sufficient condition for an $M$-affine function $g: I{\rightarrow}I$ to be $f$-affine is that $I$ is a subset of a quadrant.

In the case where $n=1$, we know more.

THEOREM 8. Let $I$ be a proper interval of $\mathbf{R}$ (not necessarily open). Let $M$ be a q.a. mean on $I$ defined by an $f: I{\rightarrow}\mathbf{R}$ which is continuous and strictly monotonic. For all $M$-affine $g: I{\rightarrow}I$ we have

$$f \circ g = af+b$$

for some real constants $a, b$, if and only if the range of $f$ is bounded from one side.

Various functional equations can be solved by using Theorem 8. We deduce for example that the general solution $g: [0,+\infty[{\rightarrow}\mathbf{R}$ of

$$\sqrt{(g(x)^2+g(y)^2)/2} = g\left(\sqrt{(x^2+y^2)/2}\right) \quad (x,y\geq 0),$$

is $\sqrt{ax^2+b}$ where $a\geq 0$, $b\geq 0$. Notice that no regularity is a priori prescribed for $g$, contrary to what happens in usual assumptions for uniqueness theorems.

We shall give here only one result for the general functional equation (12), and refer to [16] for further results and for the proof of Proposition 6, Theorem 7 and Theorem 8.

PROPOSITION 9. Let $M$ and $A$ be two q.a. means on a set $I$ generated by $f_M \colon I \to J_M$ and $f_A \colon I \to J_A$ respectively. Let $g \colon I \to I$ be a bijection. Then $g$ satisfies (12) if and only if $f_A \circ g \sim f_M$.

With the aid of such results, at least two kinds of functional equations can be solved. First, let $M$ be a q.a. mean on $I$ and look for functions $F \colon I \to J$ such that

$$F(M(x,y)) = \frac{F(x)+F(y)}{2} \qquad (x,y \in I). \tag{14}$$

To be brief, we only give an example. We take $I=]0,1[$, $M(x,y)=\sqrt{xy}$ and $J=]0,\infty[$. Then the general solution of (14) is

$$F(x) = a \log x + b; \qquad a \le 0, \; b > 0.$$

Second, let $\phi \colon I \to I$. We look for functions $g \colon I \to I$ such that

$$g\left(\frac{\phi(x)+\phi(y)}{2}\right) = \frac{1}{2}(\phi(g(x))+\phi(g(y))) \qquad (x,y \in I). \tag{15}$$

The case where $\phi$ is continuous and monotonic on an interval $I$ of $\mathbf{R}$ is not difficult. The fixed points of $\phi$ play some part here but we have no room to give detailed results. One easily finds for example that the following result holds with $I=\mathbf{R}$, $\phi(x)=x^{2n}$ and a given positive integer $n$ (cf. [17]).

PROPOSITION 10. The only solutions of (15) besides $g(x) \equiv x$ are $g(x) \equiv 0$ and $g(x) \equiv 1$.

Notice that also here there were no regularity assumption made for $g$.

It is a leitmotif of this paper to look for similar theorems concerning sets $I$ other than $\mathbf{R}$ or intervals of $\mathbf{R}$ and for replacing the hypothesis of monotonicity for $\phi$.

Instead of $M$-affine functions, one could look for $M$-convex functions, that is for functions $g\colon I \to I$ such that

$$g(M(x,y)) \leq M(g(x),g(y)), \quad (x,y \in I).$$

But we have no room here for explicit results.

Quasi-arithmetic means have been extended in various directions. The following seems to include a lot of previous constructions and to introduce an interesting type.

Let $I$ be a nonempty set and $D$ some given function defined on $I \times I$ and taking its values in a real linear space $E$. It may happen that, for any $x,y$ in $I$, the equation $D(x,z)+D(y,z)=0$ possesses a unique solution $z$ in $I$. We shall call this solution $z$ a $D$-mean of $x$ and $y$ and we write

$$z = M_D(x,y).$$

If we were to choose $D(x,y)=f(x)-f(y)$ with a bijection $f\colon I \to J$, where $J$ is a convex subset of some real linear space $E$, then $M_D(x,y)$ reduces to the quasi-arithmetic $M(x,y)$ which we defined via equation (5). When $I$ is an interval of $\mathbf{R}$, it is reasonable to require an 'internal' property concerning $M_D(x,y)$, namely for all $x,y$ in $I$,

$$\mathrm{Inf}(x,y) \leq M_D(x,y) \leq \mathrm{Sup}(x,y). \tag{17}$$

A way to get a $D$-mean satisfying (17) when $I$ is an interval of $\mathbf{R}$, is to choose a rather particular $D$, a so-called *deviation*. A function $D\colon I \times I \to \mathbf{R}$ is called a deviation if the two following properties are satisfied:

- $D(x,x)=0$ for all $x$ in $I$.

- $D$ is strictly decreasing and continuous in the second variable for each choice of the first variable.

The comparison of two $D$-means coming from two deviations $D_1$ and $D_2$ has been recently obtained in [18]. We quote here the result.

THEOREM 11. Let $D_1$ and $D_2$ be two deviations defined on an open interval $I$ of $\mathbf{R}$. To get for all $x,y$ in $I$

$$M_{D_1}(x,y) \leq M_{D_2}(x,y)$$

it is necessary and sufficient that, for all $x,y,z$ in $I$ such that $x \leq y \leq z$, we have

$$D_1(x,y)D_2(z,y) \leq D_1(z,y)D_2(x,y). \tag{18}$$

In particular, by putting an equality sign in (18), we get a necessary and sufficient condition for two deviations $D_1$ and $D_2$ to provide the same $D$-mean. Thus we may get again a uniqueness theorem for a q.a. mean, but still under a regularity assumption.

An almost untouched problem, for a given deviation $D$, is to determine all $M_D$-*affine functions*, i.e. functions $g: I \to I$ for which

$$M_D(g(x),g(y)) = g(M_D(x,y)), \quad (x,y \in I) \tag{19}$$

or $M_D$-*convex functions*, i.e. functions $g: I \to I$ for which

$$g(M_D(x,y)) \leq M_D(g(x),g(y)), \quad (x,y \in I). \tag{20}$$

Let us give two more examples of the way means can be used in order to characterize some functions. Let $M$ be a quasi-arithmetic mean on a set $I$ and let $P: I \times I \to I$. We look for $g: I \to K$, where $K$ is a commutative field, such that, for all $x,y$ in $I$,

$$g(x)g(y) = g(M(x,y))g(P(x,y)). \tag{21}$$

This functional equation was completely solved, without any regularity condition, in the case where $I = \mathbb{R}$, $K = \mathbb{R}$ and

$$M(x,y) = \frac{x+y}{2}, \quad P(x,y) = xy.$$

See [19]. From the result, we can deduce an amusing characterization of the *Heaviside function*. In fact *it is the only right-continuous solution* $g: \mathbb{R} \to \mathbb{R}$ of

$$g(x)g(y) = g(\frac{x+y}{2})g(xy) \quad (x,y \in \mathbb{R})$$

*whose range contains* $0$ *and* $1$. See [5].

However, no general treatment of (21) is available at the moment and only partial results were obtained in the case where $I$ is a topological space, $P$ has the properties (7), (8) and (9) used in the definition of quasi-arithmetic means and the mapping $I \times I \to I \times I$

$$(x,y) \to (M(x,y), P(x,y))$$

is one-to-one up to a symmetry in $x$ and $y$ (i.e. only $(x,y)$ and $(y,x)$ have the same image).

Let us give, without proof, another result. Naturally, we say that a q.a. mean on a topological space $I$ is continuous when $E_f$ is a topological linear space and $f \colon I \to J$ is a homeomorphism.

THEOREM 12. Let $I$ be a nonempty open and convex subset of $\mathbf{R}^n$. Let $M$ be a continuous q.a. mean on $I$. A continuous $g \colon \mathbf{R}^n \to \mathbf{R}^m$ is a solution of

$$g(M(x,y)) + g(x+y) = g(M(x,y) + x) + g(y), \quad (x,y \in I),$$

if and only if there exist an $m \times n$ matrix $[A]$ and a vector $a$ of $\mathbf{R}^m$ such that

$$g(x) = [A]x + a, \quad (x \in I).$$

Quasi-arithmetic means are discrete means. The continuous analogue uses integrals. An important case is that of the so-called averaging operators. To be exact, let $X$ be a compact Hausdorff space and let $\Gamma$ be a closed equivalence relation defined over $X$. Let $\mu_x$ be, for any $x$ in $X$, a positive Radon measure on $X$, of norm 1, so that for every $f$ in $C(X)$ the mapping

$$x \to \int_X f(y) \, d\mu_x(y)$$

is continuous. Here $C(X)$ is the Banach algebra of all complex-valued continuous functions over $X$ equipped with the uniform norm

$$\|f\| = \underset{x \in X}{Sup} |f(x)|.$$

Moreover we suppose that for $x\Gamma x'$ we get $\mu_x = \mu_{x'}$. For an $f$ in $C(X)$, we define $g = Pf$ by

$$g(x) = \int_X f(y) d\mu_x(y).$$

The mapping $f \to Pf$, from $C(X)$ into $C(X)$, defines an *averaging operator*.

Among all linear operators from $C(X)$ into $C(X)$ which are bounded, of norm 1 and for which $P(1)=1$, averaging operators can be characterized by the following functional equation:

$$P(f.Pg) = Pf.Pg \quad \text{for all } f,g \text{ in } C(X). \tag{22}$$

Linear operators satisfying the functional equation (22) have proved to be quite useful, not only in a Banach algebra like $C(X)$, but in other structures like the Lebesgue spaces $L_p$, in probability theory, etc. Various problems were considered in these different settings: existence of an averaging operator with a given range, integral representation of averaging operators, semi-groups of averaging operators and convergence properties etc. Some important facts in the geometry of Banach spaces are aptly described using averaging operators. Applications exist in nonlinear prediction theory, approximation theory, von Neumann algebras, mathematical logic. We cannot reproduce here the results because it would require too many explanations (for recent results, see the references in [21] and [20]). But it should be mentioned here, in the environment of this conference, that the introduction of (22) came through the practical study of turbulent fluid motion. In a turbulent fluid, the velocity of a particle or the pressure at a given point presents irregular fluctuations around an average value. It appears that averages, and averages which are not constant functions, are essential. An idea, originated by the engineer O. Reynolds at the end of the last century, is to study linear operators (with averaging properties) commuting with the Navier-Stokes differential operator. Thence came the functional equation

$$P(f.Pg+g.Pf) = Pf.Pg+P(Pf.Pg). \tag{23}$$

In general, (23) does not reduce to (22), even though this is true in many interesting circumstances. J. Kampé de Fériet, while studying (23) and (22) for turbulence theory, directed the attention of G. Birkhoff around 1950 to the algebraic properties of (23). Later G.C. Rota, around 1964, worked on (23) for $L_p$-spaces. Since then, there have been a lot of results, in algebra, in measure theory and in functional analysis. A number of similar functional equations for operators were investigated (see [22] for a survey up to 1970). We may mention *interpolation operators* which satisfy

$$P(f.Pg) = P(fg), \tag{24}$$

or *Baxter operators*, first investigated with success in queuing theory, which satisfy the functional equation

$$P(f.Pg + g.Pf) = Pf.Pg + P(fg). \tag{25}$$

Such operators led to the consideration of rather new and interesting equations for which, in various settings, general solutions were found, see [29]. We just mention that, when $(G, *)$ is an abelian group, the general solution $f: G \to G$ of the functional equation, analogous to (22),

$$f(x * f(y)) = f(x) * f(y),$$

is known. However, very little is known in the non-commutative case for the generalization $f(x * f(y)) = f(f(x) * y)$ of this equation.

## 3. Functional equations and group theory

It is always important to determine all subgroups of a given continuous group of transformations. The classical method for looking at one parameter subsemigroups presupposes strong regularity. A method using functional equations can bring some insight under weaker regularity assumptions. Let $F$ be a real topological linear space. Let us take for example the group $G$ of all couples $(\alpha, \beta)$ where $\alpha \in \mathbb{R} \setminus \{0\}$, $\beta \in F$, under the binary operation

$$(\alpha, \beta) \cdot (\alpha', \beta') = (\alpha\alpha', \alpha\beta' + \beta). \tag{26}$$

This $G$ is isomorphic to the group of proper affine transformations of the real line when $F$ coincides with $\mathbb{R}$. We keep this terminology for a real topological linear space $F$.

Let $H$ be a subset of $G$. We will say for short that $H$ has a *faithful and continuous parametrization* if there exist a topological space $E$ and a continuous mapping $g$ from $E$ onto $H$, written as

$$g(u) = (\alpha(u), \beta(u)),$$

such that either

(i)   $\beta(E)=F$, $\beta(u)=\beta(u')$ implies $\alpha(u)=\alpha(u')$, and

(ii)  $\beta$ admits *locally a continuous lifting*, i.e. for every $x$ in $F$, there exists an open neighbourhood $V$ of $x$ and a continuous $\gamma: V \to E$ such that $\beta \cdot \gamma$ is the identity function on $V$,

or

(iii) $\alpha(E)=\mathbb{R}\backslash\{0\}$, $\alpha(u)=\alpha(u')$ implies $\beta(u)=\beta(u')$, and

(iv)  $\alpha$ admits locally a continuous lifting.

What we have in mind is to look for such subsemigroups $H$ of $G$, i.e., $H \cdot H \subset H$, which have a faithful and continuous parametrization. When $E=F=\mathbb{R}$, the family $H$ of all translations is an example of such a (semi)group, in fact the only one for which case (i) and (ii) applies. Could we expect to get more such subsemigroups if we enlarge the set $E$ of parameters and if we take $F$ to be a real topological linear space? The answer is no as can be seen from the following theorem (see [5]).

THEOREM 13. Let $F$ be a real topological linear space and let $G$ be the group of all proper affine transformations from $F$ onto $F$. The only subsemigroups of $G$ which have a faithful and continuous parametrization are the group of all translations

$$x \to x+\beta \quad (\beta \in F)$$

and the groups $H_a$ ($a \in F$) defined by

$$x \to \alpha(x+a)-a \quad (\alpha \in \mathbb{R}\backslash\{0\}).$$

The technique used for the proof is interesting. We will consider only the case (i), (ii), in order to get translations. The proof is similar for the case (iii), (iv), leading to groups $H_a$.

For $u$ and $v$ in $E$, since $H$ is a subsemigroup, there exists a $w$ in $E$ such that

$$\alpha(w) = \alpha(u)\alpha(v)$$

and

$$\beta(w) = \alpha(u)\beta(v)+\beta(u).$$

The relation $f(\beta(x))=\alpha(x)$ defines unambiguously a function $f: F \rightarrow \mathbb{R}$ since $\beta(E)=F$ because $\beta(x)=\beta(x')$ implies $\alpha(x)=\alpha(x')$. Since $g$ is continuous from $E$ onto $H$ and since $\beta$ possesses locally a continuous lifting, this function $f$ is continuous. From the definition of $f$ we get

$$f(\alpha(u)\beta(v)+\beta(u)) = \alpha(w) = \alpha(u)\alpha(v) = f(\beta(u))f(\beta(v)).$$

With $x=\beta(u)$ and $y=\beta(v)$, we obtain for all $x,y$ in $F$ the functional equation

$$f(x+f(x)y) = f(x)f(y). \tag{27}$$

We can solve this functional equation when $f$ is supposed to be continuous, which is actually the case. We omit the details of the proof. The following functions are the only continuous solutions of (27):

$$f(x) = 0, \quad f(x) = 1+<x,x^*>$$

where $x^*$ is an element of the topological dual $F^*$ of $F$, and

$$f(x) = Sup(1+<x,x^*>,0).$$

But in our case, $f(x)$ must never be equal to zero. Thus we deduce that $x^*=0$ and $f(x)\equiv 1$. This form yields $\alpha(u)=1$ for all possible $u$ and therefore $H$ coincides with the family of all translations in $F$, i.e. the set of all transformations from $F$ onto $F$, $x \rightarrow x+\beta$, where $\beta \in F$.

However, the general solution $f: F \rightarrow \mathbb{R}$ of the functional equation (27) can be found. We can prove a generalization of Theorem 13 by deleting all topological properties in the parametrization we used for $H$. We then define a *faithful parametrization* with a set $E$ to be a mapping $g\colon E{\to}H$ which is onto and either (i) or (iii). If $F$ is a linear space over a commutative field, a proper affine transformation from $F$ onto $F$ has the form $x{\to}\alpha x+\beta$, $\alpha\in K\backslash\{0\}$ and $\beta\in F$.

THEOREM 14. Let $F$ be a linear space over a commutative field $K$. Let $G$ be the group of all proper affine transformations from $F$ onto $F$. The only subsemigroups of $G$ which admit a faithful parametrization are the group of all translations $(x{\to}x+\beta$, $\beta\in F)$ and the groups $H_a$ $(x{\to}\alpha(x+a)-a$, $\alpha\in K\backslash\{0\})$ where $a\in F$.

It could be interesting to replace the actual group $G$ of proper affine transformations of $F$ by the group of all transformations

$$x \to [\alpha]x+\beta$$

where $\beta\in F$ and $[\alpha]$ is an invertible and bounded linear operator from $F$ onto $F$ $([\alpha]\in GL(F))$ in the case where $F$ is a real Banach space. In this situation, the same functional equation (27) occurs, but $f(x)$ is no longer a real number but an element of $GL(F)$ and where $f(x)y$ denotes the action of the operator $f(x)$ over the vector $y$. Apparently, this generalization of the functional equation (27) has resisted a complete classification of its continuous solutions, except in the case where $F{=}\mathbf{R}^2$. See [27].

A typical feature of the functional equation (27) is that it contains *a superposition of the unknown function f*, i.e. the appearance of a term like $f(x+f(x)y)$. Functional equations with such terms are in general difficult to solve (see [20] to get an almost exhaustive list of the cases actually known). Such equations appear quite frequently for the study of subgroups. Another example is the following. We use the binary law for couples $(\alpha,\beta)$ when $\alpha\in \mathbf{R}\backslash\{0\}$ and $\beta\in \mathbf{R}$

$$(\alpha,\beta) \cdot (\alpha',\beta') = (\alpha\alpha',\alpha\beta' +\alpha' \beta).$$

$G=(\mathbb{R}\backslash\{0\})\times\mathbb{R}$ equipped with this law $\cdot$ is called the Clifford group of $\mathbb{R}$ and can be visualized as the multiplicative group of all matrices

$$\begin{bmatrix} \alpha & 0 \\ \beta & \alpha \end{bmatrix}.$$

We look for the subsemigroups of $(G, \cdot)$ which have a faithful and continuous parametrization. In the same way as previously, two functional equations appear:

$$f(f(x)y+f(y)x) = f(x)f(y) \tag{28}$$

and the classical functional equation of derivation

$$f(xy) = xf(y)+f(x)y. \tag{29}$$

Using the continuous solution of a generalization of Euler's equation (see [20]), N. Brillouët has shown that all continuous $f\colon \mathbb{R}\to\mathbb{R}$, which are solutions of (28), are constant: $f\equiv 0$, or $f\equiv 1$ (see [28].) It is then possible to state the following (see [5]).


THEOREM 15. Let $G$ be the Clifford group of $\mathbb{R}$. The only subsemigroups of $G$ which have a faithful and continuous parametrization are the group of all matrices

$$\begin{bmatrix} 1 & 0 \\ \beta & 1 \end{bmatrix}$$

with $\beta\in\mathbb{R}$ and the groups $H_a$ ($a\in\mathbb{R}$) of all matrices of the form

$$\begin{bmatrix} \alpha & 0 \\ a\alpha \log|\alpha| & \alpha \end{bmatrix} \text{ with } \alpha\in\mathbb{R}\backslash\{0\}.$$


The problem of finding the general solution of (28) is an open problem at the moment.

The method of parametrization we have used in order to find subsemigroups (which in fact proved to be subgroups) requires that such subsemigroups lie on curves. To be more precise, in the example of the Clifford group of $\mathbf{R}$, we looked for subsemigroups which can be found in the graph of a function $\alpha \to f(\alpha)$ or $\beta \to f(\beta)$ in the cartesian product of all couples $(\alpha, \beta)$ where $\alpha \in \mathbf{R} \backslash \{0\}$ and $\beta \in \mathbf{R}$. (See the figure for the two cases we found.) The problem remains open to find subsemigroups which lie in general cartesian curves

$$F(\alpha, \beta) = 0.$$

This problem requires a new technique in order to solve functional equations of a new kind, namely to find $F: (\mathbf{R} \backslash \{0\}) \times \mathbf{R} \to \mathbf{R}$ such that, as soon as $F(\alpha, \beta) = 0$ and $F(\alpha', \beta') = 0$, we get $F(\alpha \alpha', \alpha \beta' + \alpha' \beta) = 0$. This may be a promising new field.

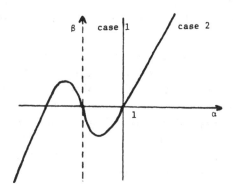

## 4. Conditional functional equations

In the investigation of functional equations arising from applications it has been frequently observed that the validity of the equation is not known for all admissible values of the variables but only under some restrictions. The restriction of the natural domain generally originates from the choice of variables which must make sense in the applied field under study. Let us sketch an example from ideal gas theory (see R. Ger in [2]). The Boltzman transport equation describing the meeting of two molecules, for vector velocities in an ideal gas, can be reduced to the following simple equation

$$h(v_1) + h(v_2) = h(v_3) + h(v_4)$$

under the double restriction

$$v_1 + v_2 = v_3 + v_4$$

and

$$\|v_1\|_2^2 + \|v_2\|_2^2 = \|v_3\|_2^2 + \|v_4\|_2^2$$

where $\|.\|_2$ denotes the euclidean norm. By a convenient change of variables and a translation on the function, we deduce the classical functional equation

$$f(x+y) = f(x) + f(y) \quad \text{for} \quad x \perp y \tag{30}$$

Here $x \perp y$ means that $x$ is orthogonal to $y$. Equation (30) is a typical *conditional functional equation*. It can be solved under a continuity assumption for $f$ (see [20]) and we thus may derive the Maxwell-Boltzman formula for the distribution of vector velocities in an ideal gas.

Even though till now only *conditional Cauchy equations* have been studied in a systematic way, the literature is already impressive and the two survey papers of 1978 (see [23] and [24]) need additions. Conditional functional equations form a very active domain in the theory of functional equations, with promising applications. It is not less interesting from a theoretical point of view, as can be seen from the following characterization of inner product spaces among normed spaces.

THEOREM 16. Let $E$ be a real normed space of dimension at least 3. We define $x \perp y$ by the James condition $\|x + \lambda y\| \geq \|x\|$ for all $\lambda$ in $\mathbf{R}$. The norm of $E$ comes from an inner product if and only if there exists $f: E \to \mathbf{R}$ which is not additive but is a solution of (30).

In this theorem, we are interested in the existence of solutions of the equation (30) which are not solutions of the same equation when the restriction on the variables $(x \perp y)$ is omitted. In other cases one is interested by a *redundant condition,* i.e. a condition which does not increase the set of solutions. We should mention here that, following a technique due to Ger, it is possible to solve conditional Cauchy equations in various algebraical structures when the restriction on the variables $x$ and $y$ is of the form

$$f(x+y) \neq af(x)+bf(y)+c.$$

(see for example [25]). Other interesting appearances of conditional Cauchy equations can be found in number theory (see [20]).

Suppose we take $y=g(x)$, where $g$ is a given function, as the restriction for the variables $x$ and $y$ in a conditional Cauchy equation. We write in a more compact form

$$f(x+g(x)) = f(x)+f(g(x)). \tag{31}$$

For a large class of $g$, the condition is redundant under strong regularity for $f$ (differentiability at zero). Here the Cauchy equation is supposed to hold only on the graph of $g$. A redundancy result for graphs has also been proved (see [30]) in the case $g=f$, again if $f$ is differentiable at 0 (and continuous on $\mathbb{R}$). With (31), we enter into the realm of functional equations in a single variable, a rich subject which unfortunately is untouched in this paper. But a long survey of M. Kuczma is supposed to appear soon, so we can conclude this survey here.

Institut de Mathématiques
Université de Nantes
2 Chemin de la Houssinière
F-44072 Nantes Cedex
France

# References

[1]   Proceedings of the Eighteenth International Symposium on Functional Equations, Aug.-Sept. 1980, Waterloo-Scarborough, Canada, Univ. of Waterloo, Ont.

[2]   Proceedings of the Nineteenth International Symposium on Functional
      Equations, May 1981, Nantes-La Turballe, France, Univ. of Waterloo, Ont.

[3]   Alexandrov, A.D.: 1967, A contribution to chronogeometry. *Canad. J.
      Math. 19,* 1119-1128.

[4]   Alexandrov, A.D.: 1971, On a certain generalization of the functional
      equation $f(x+y)=f(x)+f(y)$. *Siberian Math. J. 11,* 198-209.

[5]   Aczél, J. and J. Dhombres: *Functional Equations in Several Variables.*
      Addison-Wesley, Reading, Mass., to appear.

[6]   Forti, G.L.: 1977, On the functional equation $f(L+x)=f(L)+f(x)$. *Istit.
      Lombardo Accad. Sci. Lett. Rend. A 8,* 296-302.

[7]   Forti, G.L.: 1978, Bounded solutions with zeros of the functional equation
      $f(L+x)=f(L)+f(x)$. *Boll. Un. Mat. Ital. (5) 15-A,* 248-256.

[8]   Forti, C.B.: 1980, Some classes of solutions of the functional equation
      $f(L+x)=f(L)+f(x)$. *Riv. Mat. Univ. Parma (4) 6,* 1-9.

[9]   Täuber, J. and N. Neuhaus: 1978, Monotone Lösungen zweier komplex -
      Funktionalgleichungen die aus der Jensenfunktionalgleichung hervorgehen.
      *Rev. Roumaine Math. Pures Appl. 23,* 299-307.

[10]  Täuber, J. and N. Neuhaus: 1978, Uber die monotonen Lösungen der auf
      Mengen verallgemeinerten Jensenfunktionalgleichung. *Rev. Roumaine
      Math. Pures Appl. 23,* 309-312.

[11]  Dhombres, J., Equations fonctionnelles ensemblistes. To appear.

[12]  Aczél, J.: 1966, *Lectures on functional equations and their applications.*
      Academic Press, New York-London.

[13]  Sigmon, K.: 1970, Cancellative medial means are arithmetic. *Duke Math. J.
      37,* 439-445.

[14]  Ng, C.T.: 1971, On uniqueness theorems of Aczél and cellular internity of
      Miller. *Aequationes Math. 7,* 132-139.

[15]  Bedingfield, N.K.: 1978, *A functional equation involving vector mean values.*
      Ph.D. Monash University, Australia.

[16]  Dhombres, J.: 1981-1982, *Moyennes.* Publications mathématiques de
      l'Université de Nantes, to appear.

[17]  Aczél, J.: 1967, General solution of "isomoment" functional equations.
      *Amer. Math. Monthly 74,* 1068-1071.

[18] Daróczy, Z. and Zs. Páles, On comparison of mean values. To appear.

[19] Benz, W.: 1979, On a conjecture of I. Fenyö. *C.R. Math. Rep. Acad. Sci. Canada 1*, 249-252.

[20] Dhombres, J.: 1979, *Some aspects of functional equations*. Lecture Notes. Chulalongkorn University.

[21] Aló, R., A. de Korvin, and C. Roberts: 1977, Averaging operators on normed Köthe spaces. *Ann. Mat. Pura Appl. (4) 112*, 33-48.

[22] Dhombres, J.: 1971, Sur les opérateurs multiplicativement liés. *Mém. Soc. Math. France, no. 27*.

[23] Dhombres, J. and R. Ger: 1978, Conditional Cauchy equations. *Glasnik Mat. 13 (33)*, 39-62.

[24] Kuczma, M.: 1978, Functional equations on restricted domains. *Aequationes Math. 18*, 1-34.

[25] Forti, G.L. and L. Paganoni: 1981, A method for solving a conditional Cauchy equation in abelian groups. *Ann. Mat. Pura Appl. (4) 127*, 79-99.

[26] The Twentieth International Symposium on Functional Equations, August 1-August 7, 1982, Oberwolfach, Germany, *Aequationes Math. 24*, 261-297.

[27] Brillouët, N., *L'équation fonctionnelle de Golab-Schinzel de* $\mathbb{R}^n$ *dans* $GL_n(\mathbb{R})$. Publications mathématiques de l'Université de Nantes, to appear.

[28] Brillouët, N.: *Equations fonctionnelles et théorie des groupes*. Publications mathématiques de l'Université de Nantes, to appear.

[29] Miller, J.B.: The Euler-MacLaurin formula for an inner derivation. *Aequationes Math.*, to appear.

[30] Forti, G.L.: 1983, On some conditional Cauchy equations. *Boll. Un. Mat. Ital. (6) 2-B*, 391-402.

A. Krapež

## Groupoids with Δ-kernels

## 1. Introduction

In this note some preliminary results of recent research done by the author are presented.

In [1], [2] and [3] a uniform technique is developed for solving a wide class of functional equations on groupoids. We shall give a quick overview of the method.

Let us consider the functional equation $L(X;F)=R(X;F)$, where $X$ indicates (object) variables and $F$ stands for some unknown functions (i.e. functional variables). The equation may also include some given functions, which we call parameters.

If we replace the unknown functions from $F$ by some specific functions (say $G$), then the terms $L(X;G)$ and $R(X;G)$ define two functions. If they happen to be equal, we say that $G$ is a solution to the equation $L(X;F)=R(X;F)$.

Let $A(E)$ be the main operation of the term $L(X;F)$ $(R(X;F))$. That is, $L(X;F)$ is of the form $A(...)$, where ... stands for an appropriate sequence of variables, parameters and brackets (and similarly for $R(X;F)$).

We construct a solution of $L(X;F)=R(X;F)$ in the following way:

- we aribitrarily choose all the functions from $F$ except $A$ and $E$;

- using kernels of parameters and chosen functions from $F$, we construct two equivalences, one for each term, which depend on the structure of the terms $L(X;F)$ and $R(X;F)$;

*J. Aczél (ed.), Functional Equations: History, Applications and Theory, 93-97.*
© *1984 by D. Reidel Publishing Company.*

- if we denote by $\tau$ the least equivalence greater than both of the mentioned equivalences, then any function $T$ with a kernel greater than (or equal to) $\tau$ defines a (class of) solution(s) to the given functional equation;

- our final choice of $A$ and $E$ (i.e. completing $G$) should force the terms $L(X;G)$ and $R(X;G)$ to be both equal to the function $T$.

Thus we obtain $A$ and $E$ such that (depending on the function $T$) they cover any differences between the left and right hand side of the equation, consequently turning $G$ into a solution.

Independently of [3], S.B. Prešić [5] has developed a similar but stronger technique for solving an even wider class of equations. More on this subject can be found in [4].

Here we restrict our attention to a class of operations (we call them groupoids with $\Delta$-kernels) which generalize projection functions. We construct an algebra which is a homomorphic image of the algebra of all groupoids with $\Delta$-kernels.

## 2. Definitions

Let $D$ be a set of all words (including the empty word $e$) over the set $\{0,1\}$.

For a given nonempty set $S$ let $\mathbf{D}=(\Delta,\Omega)$ be an algebra consisting of a set of equivalences $\Delta$ and a set $\Omega$ of operations on $\Delta$. $\Delta$ and $\Omega$ are defined as follows:

- $\Delta=\{\Delta_w|w\in D\}$
- $\Omega=\{\omega_w|w\in D\}$
- If $w=p_1...p_n$ ($p_i\in\{0,1\}$) then $\Delta_w$ is the equivalence relation on $S^n$ given by $(x_1,...,x_n)\Delta_w(y_1,...,y_n)$ iff $x_i=y_i$ for all $i$ such that $p_i=1$
- $\omega_e=\Delta_e$

- $\omega_0(\Delta_w)=\Delta_{0^{|w|}}$ where $0^{|w|}$ is a word of the length $|w|$ (the length of $w$) consisting of zeros only

- $\omega_{1^n}(\Delta_{w_1},\dots,\Delta_{w_n}) = \Delta_{w_1,\dots,w_n}$ and in particular $\omega_1(\Delta_w)=\Delta_w$.

- If $w=p_1\dots p_n$ then

$$\omega_w(\Delta_{w_1},\dots,\Delta_{w_n})=\omega_{1^n}(\omega_{p_1}(\Delta_{w_1}),\dots,\omega_{p_n}(\Delta_{w_n})).$$

For example $\Delta_e$ is the equivalence on the set $\{\emptyset\}$, $\Delta_0$ is the universal relation on $S$, $\Delta_1$ is the equality on $S$ etc.

Also $\omega_e=\Delta_e$, $\omega_0(\Delta_w)=\Delta_{0^{|w|}}$, $\omega_1$ is the identity permutation of $\Delta$, $\omega_{11}(\Delta_u,\Delta_v)=\Delta_{uv}$ and so on.

## 3. Groupoids with $\Delta$-kernels and solutions of functional equations

The system $\mathbf{D}=(\Delta,\Omega)$ can serve to describe the behaviour of so-called groupoids with $\Delta$-kernels.

Any $n$-ary groupoid $P$ on $S$, such that $\ker P\in\Delta$, is a groupoid with a $\Delta$-kernel.

For example, groupoids with the kernel $\Delta_e$ are constants from $S$, groupoids with kernel $\Delta_0$ are constant mappings from $S$ to $S$, groupoids with kernel $\Delta_1$ are permutations on $S$, groupoids with kernel $\Delta_{01}$ are operations of the form $P(x,y)=\pi y$ where $\pi$ is a permutation on $S$, etc.

The main property of such groupoids is the following. If $P_0$, $P_1,\dots,P_n$ are groupoids with $\Delta$-kernels, then the groupoid

$$P(x_{11},\dots,x_{1m_1},\dots,x_{n1},\dots,x_{nm_n})$$
$$= P_0(P_1(x_{11},\dots,x_{1m_1}),\dots,P_n(x_{n1},\dots,x_{nm_n}))$$

is also a groupoid with $\Delta$-kernel.

As a consequence, the system $\mathbf{D}$ is a homomorphic image of the algebra of all groupoids with $\Delta$-kernels on $S$.

Another important property that these groupoids share is that each has a retract with the equality as kernel.

Using results from [1], [2] and [3], we can easily construct solutions of functional equations for groupoids with $\Delta$-kernels.

We replace all functions from $F$, except $A$ and $E$, by some arbitrary groupoids with $\Delta$-kernels. Since the parameters are also groupoids with $\Delta$-kernels, the resulting equivalences for the left and the right hand side of the equation belong to $\Delta$ and so does $\tau$.

Any groupoid $T$ with a $\Delta$-kernel greater than or equal to $\tau$ gives rise to possible groupoids $A$ and $E$ (they should also be groupoids with $\Delta$-kernels), which completes the construction of the solution.

We can see that $\Delta^n = \{\Delta_w \| |w| = n\}$ with appropriate operations is a Boolean algebra isomorphic to the Boolean algebra $2^n$. The zero and unit of this algebra are the universal relation and equality on $S^n$, respectively.

Using the lattice of all equivalences on a set $S$, instead of just $\Delta_0$ and $\Delta_1$, we can generalize our results to the case of groupoids with kernels made from equivalences on $S$ in the same way that groupoids with $\Delta$-kernels have kernels made from $\Delta_0$ and $\Delta_1$.

Matematički Institut
Knez Milhailova 35
Y-11000 Beograd
Yugoslavia

## References

[1] Krapež, A.: 1980, Generalized associativity on groupoids. *Publ. Inst. Math. N.S. 28 (42)*, 105-112.

[2] Krapež, A.: 1981, Functional equations of generalized associativity, bisymmetry, transitivity, and distributivity. *Publ. Inst. Math. N.S., 30 (44)*, 81-87.

[3] Krapež, A.: 1982, Generalized balanced functional equations on $n$-ary groupoids. *Proceedings of the symposium n-ary structures*, Skopje, 13-16.

[4] Krapež, A., Functional equations on groupoids and related structures. This volume.

[5]  Prešić, S.B.: 1982, A general solving method for one class of functional equations. *Proceedings of the symposium n-ary structures*, Skopje, 1-9.

A. Tsutsumi and Sh. Haruki

### The regularity of solutions of functional equations and hypoellipticity

## 1. Introduction

In this paper we investigate the regularity of solutions of functional equations of the form

$$\sum_{j=1}^{k} a_j(x,t) f(h_j(x,t))$$

$$= F(x, f(l_1(x)),...,f(l_s(x))) + b(x,t), \tag{1.1}$$

where

$$x \in \mathbf{R}^n, \quad t \in \omega \subseteq \mathbf{R}^r, \quad n > 1, \; r \geq 1,$$

$$h_j: \mathbf{R}^n \times \omega \rightarrow \mathbf{R}^n, \quad l_j: \mathbf{R}^n \rightarrow \mathbf{R}^n,$$

$$a_j: \mathbf{R}^n \times \omega \rightarrow \mathbf{R}, \quad b: \mathbf{R}^n \times \omega \rightarrow \mathbf{R},$$

$$F: \mathbf{R}^{n+s} \rightarrow \mathbf{R},$$

in particular, the problem of whether all continuous or all locally integrable solutions of (1.1) are $C^\infty$. An affirmative answer makes it easier to find all such solutions of (1.1) because we can then use a powerful method: differentiation.

For the case of $n = r = 1$ a similar result was obtained by a method of classical analysis (see Aczél [1], pp. 183-209). For the general case of $n > 1$, $r \geq 1$, methods of classical analysis do not work. Therefore generalized solutions of (1.1) in the sense of distributions are considered.

The main idea for proving the regularity of generalized solutions of (1.1) (cf. Swiatak [15]) is the following. Consider (1.1) as an equation between two distributions and differentiate it with respect to the parameter $t$ as often as necessary in order to obtain at a

99

J. Aczél (ed.), Functional Equations: History, Applications and Theory, 99-112.
© 1984 by D. Reidel Publishing Company.

point $t=t^0$ a partial differential equation of a certain special type. These types of partial differential equation may happen to be "elliptic" or more generally "hypoelliptic". Roughly speaking all the distribution solutions of these types of partial differential equation are $C^\infty$ functions. Thus the feedback of these results causes the regularity of the solutions of (1.1).

These methods have been applied by several authors: Friedman-Littman [5] treated the equation of the type

$$u(x) = \int_K u(x+ty)\,d\mu(y).$$

Garsia [6] has refined the result of [5]. Flatto [4] treated the equation of the type

$$\int_K u(x+ty)\,d\mu(y) = 0.$$

Observe that (1.1) is a special case of these equations when the measure $\mu(y)$ is discrete. Aczél-Haruki-McKiernan-Saković [2] treated the Haruki equation

$$4f(x_1,x_2)-f(x_1+t,x_2+t)-f(x_1-t,x_2+t)$$

$$-f(x_1+t,x_2-t)-f(x_1-t,x_2-t) = 0.$$

Differentiation twice with respect to $t$ of the above equation leads to a partial differential equations of elliptic type. Swiatak [5] treated equations of the type (1.1) which she transformed to non-elliptic hypoelliptic partial differential equations with constant coefficients and hypoelliptic of constant strength (see Section 3). Swiatak's work suggests that the existence of more general hypoelliptic partial differential equations may imply the regularity of solutions of (1.1) under weaker assumptions. In fact many criteria for hypoellipticity have been investigated recently. These results will be listed in Section 3. Notations and definitions will be stated in Section 2. The main theorem will be stated and proved in Section 3. In Section 4 we will illustrate by examples that the main theorem contains new results.

## 2. Notations and definitions

Let

$$x, y, \xi, \eta \in \mathbf{R}^n, \quad t \in \mathbf{R}^r.$$

Let

$$\alpha = (\alpha_1, \ldots, \alpha_n), \quad \beta = (\beta_1, \ldots, \beta_r)$$

be nonnegative multi-integers $(\alpha_1, \ldots, \alpha_n, \beta_1, \ldots, \beta_r$ are nonnegative integers),

$$|\alpha| = \alpha_1 + \ldots + \alpha_n,$$

$$|\beta| = \beta_1 + \ldots + \beta_r,$$

$$\xi^\alpha = \xi_1^{\alpha_1} \ldots \xi_n^{\alpha_n},$$

$$\partial_x^\alpha = \partial_{x_1}^{\alpha_1} \ldots \partial_{x_n}^{\alpha_n}, \quad \partial_{x_j}^\alpha = \partial^\alpha / \partial x_j^\alpha,$$

$$D_x^\alpha = (-i)^{|\alpha|} \partial_x^\alpha,$$

$\partial_t^\beta$ and $D_t^\beta$ are similarly defined, $supp \; \phi = \{x: \phi(x) \neq 0\}$, $\mathbf{D}$ is the space of $C^\infty$ functions of compact support with the pseudo-topology in the sense of Schwartz. $\mathbf{D}'$ is the set of distributions, that is, the dual of $\mathbf{D}$ with respect to the pseudo-topology.

A distribution $T \in \mathbf{D}'$ is said to be a function of class $C$ or $L_{loc}^1$ if there exists a function $f$ in $C$ or in $L_{loc}^1$ such that

$$(T, \phi) = \int_{\mathbf{R}^n} f(x) \phi(x) dx, \quad \phi \in \mathbf{D}.$$

Distributions with parameter are also considered. Let $\omega \subseteq \mathbf{R}^r$, and $\psi: \mathbf{R}^n \times \omega \to \mathbf{R}$ and $\psi \in \mathbf{D}$. Then, for each fixed $t \in \omega$, $\Psi_t(x) = \psi(x, t)$ defines a function $\Psi_t: \mathbf{R}^n \to \mathbf{R}$ which is in $\mathbf{D}$. In analogy to the above, if $\Psi_t \in \mathbf{D}$ and $T \in \mathbf{D}'$, we may define

$$\gamma(t) = (T, \psi_t) = (T(x), \psi(x, t))_x.$$

Differentiability of $\gamma$ with respect to $t$ can be derived under suitable conditions on $\psi$ (see [15], Lemma 4.1, p. 102). For further results about distributions the reader is referred to Schwartz [14].

Further notations:

$$h_j(x, t) = (h_{j1}(x, t), \ldots, h_{jn}(x, t)),$$

$$\partial_t^\beta h_j(x,t) = (\partial_t^\beta h_{j1}(x,t),...,\partial_t^\beta h_{jn}(x,t)),$$

$$h_j: \mathbf{R}^n \times \omega \to \mathbf{R}^n, \quad h_{j1}: \mathbf{R}^n \times \omega \to \mathbf{R} \quad (i=1,...,n; j=1,...,k).$$

We say that $h_j \in C^p$ on $\mathbf{R}^n \times \omega$, where $\omega$ is an open set contained in $\mathbf{R}^r$, if $h_{ji} \in C^p$ on $\mathbf{R}^n \times \omega$, for $i=1,...,n$. We shall often write $h_j(x,t) \in C^p$ instead of $h_j \in C^p$. We write $h_j(x,t) \in C^{\bar{p}}$ on $\mathbf{R}^n$ for each fixed $t \in \omega$ if the functions $h_{ji}: \mathbf{R}^n \to \mathbf{R}^n$, defined by $h_{ji}: x \to h_{ji}(x,t)$, are functions of class $C^{\bar{p}}$ on $\mathbf{R}^n$ for all values of the parameter $t \in \omega$. If $a: \mathbf{R}^n \times \omega \to \mathbf{R}$ and if the functions $a_t(x)=a(x,t)$ are of class $C^{\bar{p}}$ on $\mathbf{R}^n$ for all values of the parameter $t \in \omega$, we write $a(x,t) \in C^{\bar{p}}$ on $\mathbf{R}^n$ for fixed $t \in \omega$.

For a linear partial differential operator of order $m$,

$$P = P(x,D_x) = \sum_{|a| \le m} a_\alpha(x) D_x^\alpha,$$

we set

$$P(x,\xi) = \sum_{|a| \le m} a_\alpha(x) \xi^\alpha$$

and

$$P_{(\beta)}^{(\alpha)}(x,\xi) = \partial_\xi^\alpha D_x^\beta P(x,\xi), \quad \xi \in \mathbf{R}^n.$$

## 3. Regularity criteria

For the equation (1.1) we suppose

(a)   $a_j(x,t), b(x,t) \in C^\infty(\mathbf{R}^n)$ $(j=1,...,k)$ for every fixed $t$ in an open set $\omega \subseteq \mathbf{R}^r$,

(b)   $a_j(x,t), b(x,t) \in C^m(\mathbf{R}^n \times \omega)$ $(j=1,...,k)$,

(c)   the mappings $x \to y = h_j(x,t)$ are diffeomorphisms on $\mathbf{R}^n$ for every fixed $t \in \omega$ $(j=1,...,k)$,

(d)   $h_j(x,t) \in C^m$ and where $h_j^{-1}$ denotes the inverse of $h_j$ with respect to $t$, $h_j^{-1}(x,t) \in C^m$ on $\mathbf{R}^n \times \omega$, $(j=1,...,k)$,

(e)   $F(x,z_1,...,z_s) \in C(\mathbf{R}^{n+s})$,

(f)   $l_j(x) \in C(\mathbf{R}^n)$, $(j=1,...,s)$.


A locally integrable function $f$ is said to be a solution of (1.1) in the sense of distributions (or in the distribution sense) if

$$\sum_{j=1}^{k} \int_{\mathbf{R}^n} a_j(x,t) f(h_j(x,t)) \phi(x) dx$$

$$= \int_{\mathbf{R}^n} F(x, f(l_1(x)),...,f(l_s(x))) \phi(x) dx$$

$$+ \int_{\mathbf{R}^n} b(x,t) \phi(x) dx \tag{3.1}$$

for each $\phi \in \mathbf{D}$ and every fixed $t \in \omega \subseteq \mathbf{R}^r$. We can write for short

$$\sum_{j=1}^{k} (a_j(x,t) f(h_j(x,t)), \phi(x))_x$$

$$= (F(x, f(l_1(x)),...,l_s(x))), \phi(x))_x$$

$$+ \int_{\mathbf{R}^n} (b(x,t), \phi(x))_x. \tag{3.1'}$$


A partial differential operator $P = P(x, D_x)$ defined on an open set $\Omega \subseteq \mathbf{R}^n$ with $C^\infty$ coefficients is said to be *hypoelliptic* on $\Omega$, if, for $u \in \mathbf{D}'(\Omega)$ and any open subset $\Omega'$ of $\Omega$, the relation $Pu \in C^\infty(\Omega')$ leads to $u \in C^\infty(\Omega')$.

The notion of hypoellipticity comes from the problem of whether or not a distribution solution of the partial differential equation $Pu = f$ is a classical solution (see Schwartz [14]). Many sufficient conditions for hypoellipticity have since been obtained. We will cite here some of the main results in the order of increasing generality.


*Hp-1.* One of the characterizations given by Hörmander [8] of hypoellipticity for a P.D.O. (partial differential operator) with constant coefficients is

$$\lim_{\xi \to \infty} P^{(\alpha)}(\xi)/P(\xi) = 0 \quad \text{for} \quad \alpha \neq 0.$$

*Hp-2.* The operator $P(x,\xi)$ is said to be of *constant strength* on an open set if, for any $x,y \in \Omega$, there exists a constant $C_{x,y}$ such that

$$C_{x,y}^{-1}\tilde{P}(x,\xi) \leq \tilde{P}(y,\xi) \leq C_{x,y}\tilde{P}(x,\xi)$$

holds, where

$$\tilde{P}(x,\xi) = \{\sum_\alpha |P^{(\alpha)}(x,\xi)^2\}.$$

When $P(x,\xi)$ is of constant strength on $\Omega$ and $P(x^0,\xi)$ is hypoelliptic as a P.D.O. with constant coefficients, $P(x,\xi)$ is said to be *formally hypoelliptic* on $\Omega$. Formally hypoelliptic operators on $\Omega$, are hypoelliptic on $\Omega$, as proved by Hörmander [9], Malgrange [13] and Trèves [16].

*Hp-3.* The operator $P(x,\xi)$ is said to satisfy the (*H*)–*condition* if there exist constants

$$C>0,\ C_{\alpha\beta}\geq 0,\ -\infty<m' <\infty,\ 0\leq\delta<\rho\leq 1$$

such that, for $x \in \Omega \subseteq \mathbf{R}^n$, $\xi \in \mathbf{R}^n$,

$$|P(x,\xi)| \geq C(1+|\xi|)^{m'}$$

for large

$$|\xi| = (\xi_1^2+...+\xi_n^2)^{\frac{1}{2}},$$

and

$$|P_{(\beta)}^{(\alpha)}(x,\xi)/P(x,\xi)| \leq P|(x,\xi)|$$
$$\leq C_{\alpha\beta}(1+|\xi|)^{-\rho|\alpha|+\delta|\beta|}$$

for large $|\xi|$. If $P(x,\xi)$ satisfies the (*H*)-condition, then $P(x,D)$ is hypoelliptic in $\Omega$. This was proved by Hörmander [10]. A modification of this result was given by Tsutsumi [17], [18].

*Hp-4.* Combining the results obtained by Kato [11], Grushin [7] and Kumano-go-Taniguchi [12], the following criterion is obtained. We write

$$m=(m_1,...,m_n),\quad \bar{m}=\max(m_j),$$

$$|\alpha:m| = \sum_{j=1}^{n} \alpha_j/m_j, \quad x=(x',x''\,),$$

$$\xi=(\xi',\xi''\,)\in \mathbf{R}^\nu \times \mathbf{R}^{n-\nu},$$

$$x=(x',\tilde{x}'',\tilde{x}''\,)\in \mathbf{R}^\nu \times \mathbf{R}^\mu \times \mathbf{R}^{n-(\nu+\mu)},$$

$(1 \leq \nu \leq \mu \leq n)$, $\gamma=(\gamma_1,\ldots,\gamma_\nu,0,\ldots,0)$, $(x',\tilde{x}''\,)^\gamma = x_1^{\gamma_1}\ldots x_\mu^{\gamma_\mu}$. Consider the P.D.O. of the form

$$P(x,D_x) = \sum a_{\alpha\gamma}(x)(x',\tilde{x}''\,)^\gamma D_x^\alpha \tag{3.2}$$

where $a_{\alpha\gamma} \in C^\infty(\Omega)$ and the summation $\sum$ is done over these $\{\alpha,\gamma\}$ which satisfy $|\alpha:m|=1$ and $(\rho,\alpha) \leq (\sigma,\gamma) \leq (\rho,\alpha)-\overline{m}$ for some fixed indices $\rho$ and $\sigma$. We also define from (3.2)

$$\mathbf{P}(x',\tilde{x}'',D_x) = \sum_0 a_{\alpha\gamma}(0)(x',\tilde{x}''\,)^\gamma D_x^\alpha, \tag{3.3}$$

where the summation $\sum_0$ is done for those $\{\alpha,\gamma\}$ which satisfy $(\rho,\alpha)=(\sigma,\gamma)-\overline{m}$.

Consider the following conditions.

C.1 There exist two multi-indices $\rho=(\rho_1,\ldots,\rho_n)$ and $\sigma=(\sigma_1,\ldots,\sigma_n)$ such that

(i) $\rho_j=\sigma_j=\overline{m}/m_j$ for $j \geq \nu$,

(ii) $\rho_j > \sigma_j \geq 0$, $m_j\rho_j \geq \overline{m}$ for $\nu < j \leq n$,

(iii) $\sigma_j=0$ for $j \geq \nu+1$,

C.2 $\mathbf{P}(\lambda^{-\sigma}(x',\tilde{x}''\,),\lambda^\rho\xi) = \lambda^{\overline{m}}\mathbf{P}(x',\tilde{x}'',\xi)$ for $\lambda>0$,

where

$$\lambda^{-\sigma}(x',\tilde{x}''\,) = (\lambda^{-\sigma_1}x_1,\ldots,\lambda^{-\sigma_\mu}x_\mu)$$

and

$$\lambda^\rho\xi = (\lambda^{\rho_1}\xi_1,\ldots,\lambda^{\rho_n}\xi_n).$$

We define from (3.3)

$$\mathbf{P}_0(x',\tilde{x}'',D_x) = \sum_{|\alpha:m|=1} a_{\alpha\gamma}(0)(x',\tilde{x}''\,)^\gamma D_x^\alpha. \tag{3.4}$$

C.3 $P_0(z' , \tilde{x}' , \xi) = 0$ for $(z' , \tilde{x}' ) \neq 0$ and $\xi \neq 0$, that is, $P_0(z' , \tilde{x}' , D_z)$ is semi-elliptic at $(z' , \tilde{x}' ) \neq 0$.

C.4 For any $\tilde{x}'$ and $\tilde{\xi}'' = (\xi_{\nu+1}, \ldots, \xi_n)$ with $|\tilde{\xi}''| = 1$ the equation $P(z' , \tilde{x}' , D_{z'} , \tilde{\xi}'' )v = 0$ does not have any nontrivial solution in $S(\mathbb{R}^\nu_{z'})$.

A sufficient condition for hypoellipticity, which we will use, is the following:

*If the operator (3.2) satisfies C.1, C.4 and*

$$\max_{\nu \le j \le n} \{\sigma_j\} = \min_{\nu < j, 1 = n} \{m_n \rho_j / m_1\},$$

*then the operator (3.2) is hypoelliptic on $\Omega$.*

The following examples illustrate this criterion.

*Example (i).* Take the operator $P = (-\Delta_{z'})^l + |z'|^{2k}(-\Delta_{z''})^m$ and suppose that $\Omega$ contains the origin. When we take

$$\rho_1 = \ldots = \rho_\nu = \sigma_1 = \ldots = \sigma_\nu = \overline{m}/l,$$

for $\overline{m} = \max\{m, l\}$ and

$$\rho_{\nu+1} = \ldots \rho_n = (k/l+1)\overline{m}/m,$$

$\sigma_{\nu+1} = \ldots = \sigma_n = 0$, we see that this operator satisfies this criterion. Thus $P$ is hypoelliptic on $\Omega$. Outside a neighbourhood of the origin $P$ is elliptic.

*(ii)* Consider the operator $P_\pm = D_{z_1} \pm i z_1^k D_{z_2}^l$ in $\mathbb{R}^2$. Take $\rho_1 = \sigma_1 = 1$, $\rho_2 = k+1$, $\sigma_2 = 0$. If $k$ is even, or $k$ is odd and $l$ is even, then $P_\pm$ is hypoelliptic. If $k$ is even, then $P_-$ is hypoelliptic.

Now we state a regularity theorem concerning the solutions of the functional equation (1.1) in a form to which we can apply the sufficient conditions for hypoellipticity cited above.

THEOREM. Suppose that the equation (1.1) satisfies the conditions (a) - (f). Moreover, suppose there exists $t^0 \in \omega$ such that $h_j(x,t^0) \equiv x$ for $j=1,...,k$, and a multi-integer $q$ $(q=(q_1,...,q_r),$ $|q| \leq m)$, such that the equation

$$\partial_t^q \left( \sum_{j=1}^k a_j(x,t) f(h_j(x,t)) \right) \Big|_{t=t^0} = 0, \qquad (3.5)$$

where $\partial_t^q$ operates formally, is hypoelliptic in $\mathbf{R}^n$.

Then every continuous solution of (1.1) is $C^\infty$, and every locally integrable solution is a function of class $C^\infty$ almost everywhere.

Proof. For the function $f$ as a distribution on $\mathbf{R}^n$ there exists a sequence $f_n \in C^\infty(\mathbf{R}^n)$ such that

$$\lim_{n \to \infty} (f_n(x), \phi(x))_x = (f(x), \phi(x))_x$$

for any $\phi \in D(\mathbf{R}^n)$. Consequently we have

$$\lim_{n \to \infty} (f_n(h_j(x,t)), \phi(x))_x = (f(h_j(x,t)), \phi(x))_x,$$

because

$$\lim_{n \to \infty} (f_n(y), \phi(h_j^{-1}(y,t)) |\det \partial h_j^{-1} / \partial y|)_y$$

$$= (f(y), \phi(h_j^{-1}(x,t) |\det \partial h_j^{-1} / \partial y|)_y$$

$$= (f(h_j(x,t)), \phi(x))_x.$$

Making use of the above, $\partial_t^q$ can be applied to both sides of (3.1'):

$$\partial_t^q \sum_{j=1}^k (a_j(x,t) f(h_j(x,t)), \phi(x))_x$$

$$= \sum_{j=1}^k \lim_{n \to \infty} \partial_t^q [(a_j(x,t) f_n(h_j(x,t)), \phi(x))_x]$$

$$= \sum_{j=1}^k \lim_{n \to \infty} (f_n(y), (-\partial_t)^q [a_j(y) \phi(h_j^{-1}(y,t)) |\det \partial h_j^{-1} / \partial y|])_y$$

$$= \sum_{j=1}^{k} (f(y),(-\partial_t)^q [a_j(y)\phi(h_j^{-1}(y,t)|\det \partial h_j^{-1}/\partial y||)_y$$

$$= \sum_{j=1}^{k} (\partial_t^q [a_j(x,t) f(h_j(x,t)), \phi(x))_x.$$

Thus we get

$$\sum_{j=1}^{k} (\partial_t^q [a_j(x,t) f(h_j(x,t))], \phi(x))_x$$

$$= (\partial_t^q b(x,t), \phi(x))_x,$$

that is, in the sense of distributions,

$$\partial_t^q (\sum_{j=1}^{k} a_j(x,t) f(h_j(x,t))) = \partial_t^q b(x,t).$$

Setting $t=t^0$, we obtain the partial differential equation

$$\partial_t^q (\sum_{j=1}^{k} a_j(x,t^0) f(x)) = \partial_t^q b(x,t^0).$$

From (a) and (b) and the hypoellipticity of the principal part of the above partial differential equation, we have that the solution $f \in C^\infty(\mathbf{R}^n)$ if $f \in C(\mathbf{R}^n)$ and that $f \in C^\infty(\mathbf{R}^n)$ almost everywhere if $f$ is locally integrable.

## 4. Applications

(1) Haruki's equation

$$4f(x_1,x_2) - f(x_1+t,x_2+t) - f(x_1-t,x_2+t)$$

$$- f(x_1+t,x_2-t) - f(x_1-t,x_2-t) = 0.$$

This is the special case of (1.1) with

$$a_1(x,t) = 4, \quad a_j(x,t) = -1, \quad (j=2,...,5),$$

$$h_1(x,t) = x, \quad h_2(x,t) = (x_1+t,x_2+t),$$

$$h_3(x,t) = (x_1-t,x_2+t), \quad h_4(x,t) = (x_1+t,x_2-t),$$

$$h_5(x,t) = (x_1-t,x_2-t),$$

$$F(x, f(l_1(x)), \ldots, f(l_s(x))) = 0, \quad b(x, t) = 0.$$

Differentiating twice with respect to $t$, we obtain

$$\Delta_x f(x) = 0 \quad \text{for} \quad t = 0.$$

The Laplacian $\Delta_x$ is elliptic. Therefore this equation is hypoelliptic from Hp-1.

(2) The equations

$$f(x_1+t, x_2) + t^2 f(x_1, x_2+t) + (t-t^2-1)f(x_1, x_2)$$
$$= t(x_1^2 + 2x_1 + t),$$
$$f(x_1+t, x_2) + f(x_1-t, x_2) + f(x_1, x_2+t^2)$$
$$+ f(x_1, x_2+t^2) - 4f(x_1, x_2) = 0.$$

Differentiating these equations four times with respect to $t$, we obtain

$$\partial_{x_1}^4 f(x_1, x_2) + l\partial_{x_2}^2 f(x_1, x_2) = 0 \quad (l > 0).$$

This is a semi-elliptic equation. Hence it is hypoelliptic from Hp-1.

(3) In [15] (Theorem 6.2, pp. 111-112), Swiatak gave a sufficient condition for the equation (3.5) to be formally hypoelliptic. Swiatak illustrated its application by an example which led to a hypoelliptic partial differential equation with constant coefficients. We were inspired by her work to study this problem.

(4) The equation

$$f(x_1-t, x_2) + f(x_1+t, x_2) + x_1^2 f(x_1, x_2-t)$$
$$+ x_1^2 f(x_1, x_2+t)$$
$$= 2f(x_1, x_2) + 2x_1^2 f(x_1, x_2) + 2[f(x_1, x_2)]^2$$
$$+ [f(x_1+x_2, x_1-x_2)]^2$$
$$- [f(x_1, x_1)]^2 - [f(x_2, x_2)]^2.$$

Differentiating twice with respect to $t$ and setting $t=0$, we obtain

$$\partial_{x_1}^2 f(x_1, x_2) + x_1^2 \partial_{x_2}^2 f(x_1, x_2) = 0.$$

The operator

$$P(x,D_x) = \partial^2_{x_1} + x_1^2 \partial^2_{x_2}$$

is hypoelliptic by Example (i) in Section 3. Notice that this example is not formally hypoelliptic as in (3). In [15], all the $a_j(x,t)$ $j=1,\ldots,k$, of the equation (1.1) are assumed to be positive. But in this example $a_3(x,t)=a_4(x,t)=x_1^2$, $a_6(x,t)=-x_1^2$ show that these $a_j(x,t)$ are degenerate at $x_1=0$.

(5) Our final example is

$$f(x_1+t,x_2)+ix_1^4 f(x_1,x_2+t^2)-2f(x_1,x_2) = 0.$$

Differentiating twice with respect to $t$ and setting $t=0$ we obtain

$$\partial_{x_1} f(x_1,x_2)+ix_1^4 \partial_{x_2} f(x_1,x_2) = 0.$$

The operator

$$P(x,D_x) = \partial_{x_1}+ix_1^4 \partial_{x_2}$$

is hypoelliptic by Example (ii) of the previous section.

By similar considerations, we may produce many other examples as applications of the main theorem.

Department of Mathematics
College of Liberal Art and Science
Okayama University
Okayama 700
Japan

Department of Applied Mathematics
Graduate School
Okayama University of Science
Okayama 700
Japan

## References

[1]     Aczél, J.: 1966, *Lectures on functional equations and their applications.* Academic Press, New York-London.

[2]  Aczél, J., H. Haruki, M.A. McKiernan and G.N. Sakovic: 1968, General and regular solutions of functional equations characterizing harmonic polynomials. *Aequationes Math. 1*, 37-53.

[3]  Fenyö, I.: 1966, Über eine Funktionalgleichung. *Math. Nachr. 81*, 103-109.

[4]  Flatto, L.: 1963, Functions with a mean value property II. *Amer. J. Math. 85*, 248-270.

[5]  Friedman, A. and W. Littman: 1962, Functions satisfying the mean value property. *Trans. Amer. Math. Soc. 102*, 167-180.

[6]  Garsia, A.M.: 1962, A note on the mean value property. *Trans Amer. Math. Soc. 102*, 181-186.

[7]  Grushin, V.V.: 1972, Hypoelliptic differential equations and pseudo-differential operators with operator-valued symbols. *Math. U.S.S.R. Sb. 17*, 497-514.

[8]  Hörmander, L.: 1955, On the theory of general partial differential operators. *Acta Math. 94*, 161-248.

[9]  Hörmander, L.: 1961, Hypoelliptic differential operators. *Ann. Inst. Fourier 11*, 477-492.

[10]  Hörmander, L.: 1967, Pseudodifferential operators and hypoelliptic equations. In: *Proc. Symp. Pure Math. Vol. 10, Amer. Math. Soc.*, Providence, RI, pp. 138-183.

[11]  Kato, Y.: 1970, On a class of hypoelliptic differential operators. *Proc. Japan Acad. Ser. A. Math. Sci. 46*, 33-37.

[12]  Kumano-go, K. and K. Taniguchi: 1973, Oscillatory integrals of symbols of pseudo-differential operators on $R^n$ and operator of Fredholm type. *Proc. Japan Acad. Ser. A. Math. Sci. 49*, 397-402.

[13]  Malgrange, B.: 1957, Sur une classe d'opérateurs différentiels hypoelliptiques. *Bull. Soc. Math. France 85*, 283-306.

[14]  Schwartz, L.: 1966, *Théorie des distributions*. Herman, Paris.

[15]  Swiatak, H.: 1976, The regularity of the locally integrable and continuous solutions of nonlinear functional equations. *Trans. Amer. Math. Soc. 221*, 97-118.

[16]  Trèves, F.: 1959, Operateurs différentiels hypoelliptiques. *Ann. Inst. Fourier 9*, 1-73.

[17]  Tsutsumi, A.: 1973, Remark on a sufficient condition for hypoellipticity. *Sci. Rep. College General Ed. Osaka Univ. 22*, 23-31.

[18]  Tsutsumi, A.: 1975, On the asymptotic behavior of resolvent kernels and spectral functions for some class of hypoelliptic operators. *J. Differential Equations 18*, 366-385.

[19]  Tsutsumi, A. and S. Haruki: 1982, Functional equations and hypoellipticity. *Proc. Japan Acad. Ser. A. Math. Sci. 58*, 105-108.

[20]  Tsutsumi, A. and S. Haruki: 1982, Functional equations and hypoellipticity. In: *Supplement to the Proc. of the 2nd World Conf. on Math. at the Service of Man*, Las Palmas, Spain.

[21]  Zalcman, L.: 1973, Mean values and differential equations. *Israel J. Math. 14*, 339-352.

H. Haruki

## An improvement of the Nevanlinna-Pólya theorem

## 1. Introduction, statement of theorem

The Nevanlinna-Pólya theorem (see [3], [4], [6]) states the following.

THEOREM A. Let $n$ be an arbitrarily fixed positive integer and let $f_k$ and $g_k$ $(k=1,2,...,n)$ be regular functions of a complex variable $z$ on a nonempty domain $D$. If $f_k$ and $g_k$ $(k=1,2,...,n)$ satisfy

$$\sum_{k=1}^{n} |f_k(z)|^2 = \sum_{k=1}^{n} |g_k(z)|^2 \tag{1}$$

on $D$ and if $f_1, f_2,...,f_n$ are linearly independent on $D$, then

(i)  there exists an $n \times n$ unitary matrix $\mathbf{C}$ where each of the elements of $\mathbf{C}$ is a complex constant such that

$$\begin{pmatrix} g_1(z) \\ g_2(z) \\ \cdot \\ \cdot \\ \cdot \\ g_n(z) \end{pmatrix} = \mathbf{C} \begin{pmatrix} f_1(z) \\ f_2(z) \\ \cdot \\ \cdot \\ \cdot \\ f_n(z) \end{pmatrix} \tag{2}$$

holds on $D$,

113

J. Aczel (ed.), Functional Equations: History, Applications and Theory, 113-126.
© 1984 by D. Reidel Publishing Company.

(ii)   $g_1, g_2, ..., g_n$ are also linearly independent on $D$.

REMARK. It is clear that, conversely, $g_1, g_2, ..., g_n$ satisfying (2) satisfy (1).

The following generalization of Theorem A has been proved (see [3]):

THEOREM B. Let $l, m$ be arbitrarily fixed positive integers satisfying $l \leq m$ and let $f_i$ $(i=1,2,...,l)$ and $g_j$ $(j=1,2,...,m)$ be regular functions of $z$ on a nonempty domain $D$. If $f_i$ $(i=1,2,...,l)$ and $g_j$ $(j=1,2,...,m)$ satisfy

$$\sum_{i=1}^{l} |f_i(z)|^2 = \sum_{j=1}^{m} |g_j(z)|^2 \tag{3}$$

on $D$ and if $f_1, f_2, ..., f_l$ are linearly independent on $D$, then

(i)    there exists a constant complex $m \times l$ matrix $\mathbf{C} = (c_{ji})$ $(j=1,2,...,m; \; i=1,2,...,l)$, satisfying

$$\sum_{j=1}^{m} c_{jp} \bar{c}_{jq} = \delta_{pq} \quad (p,q=1,2,...,l),$$

(where $\delta_{pq}$ is the "Kronecker delta"), such that

$$\begin{pmatrix} g_1(z) \\ g_2(z) \\ \cdot \\ \cdot \\ \cdot \\ g_m(z) \end{pmatrix} = \mathbf{C} \begin{pmatrix} f_1(z) \\ f_2(z) \\ \cdot \\ \cdot \\ \cdot \\ f_l(z) \end{pmatrix} \tag{4}$$

holds on $D$,

(ii)   $g_1, g_2, ..., g_m$ are linearly independent or linearly dependent according to whether $l = m$ or $l < m$.

REMARK. It is clear that $g_1, g_2, ..., g_m$ satisfying (4) satisfy also (3).


The purpose of this note is to prove the following theorem which is a generalization of Theorem A and Theorem B.


THEOREM. Let $l, m, s$ be arbitrarily fixed positive integers satisfying $s \leq l \leq m$ and let $f_i$ $(i=1,2,...,l)$ and $g_j$ $(j=1,2,...,m)$, be regular functions of $z$ on a nonempty domain $D$. If $f_i$ $(i=1,2,...,l)$ and $g_j$ $(j=1,2,...,m)$ satisfy the functional equation (3) on $D$, if the dimension of the complex vector space spanned by the set $\{f_1, f_2, ..., f_l\}$ is $s$, and if $f_{n_1}, f_{n_2}, ..., f_{n_s}$ $(n_1 < n_2 < ... < n_s)$ are linearly independent on $D$, then

(i)  there  exist  complex  coefficients  $a_{pi}$ $(i=1,2,...,s;$ $p \notin S = \{n_1, n_2, ..., n_s\})$, such that

$$f_p(z) = \sum_{i=1}^{s} a_{pi} f_{n_i}(z) \quad (p \notin S) \tag{5}$$

holds on $D$;

(ii)  there  exists  a  constant  complex  $m \times s$  matrix  $C = (c_{ji})$ $(j=1,2,...,m;\ i=1,2,...,s)$ satisfying

$$\sum_{p \notin S} |a_{pi}|^2 + 1 = \sum_{j=1}^{m} |c_{ji}|^2 \quad (i=1,2,...,s) \tag{6}$$

and

$$\sum_{p \notin S} a_{pi} \bar{a}_{pk} = \sum_{j=1}^{m} c_{ji} \bar{c}_{jk} \quad (i,k=1,2,...,s; i \neq k) \tag{7}$$

such that

$$\begin{pmatrix} g_1(z) \\ g_2(z) \\ \cdot \\ \cdot \\ \cdot \\ g_m(z) \end{pmatrix} = C \begin{pmatrix} f_{n_1}(z) \\ f_{n_2}(z) \\ \cdot \\ \cdot \\ \cdot \\ f_{n_s}(z) \end{pmatrix} \tag{8}$$

holds on $D$.

REMARK 1. When $s=l$, we adopt the conventions that $\sum_{p\notin S}|a_{pi}|^2$ and $\sum_{p\notin S}a_{pi}\bar{a}_{pk}$ mean 0.

REMARK 2. It is clear that, conversely, $f_p,(p\notin S),g_1,g_2,...,g_m$ satisfying (5) and (8) satisfying (3).

## 2. Proof of the theorem

Proof of (i): The proof is clear from the fact that the set $\{f_{n_1},f_{n_2},...f_{n_s}\}$ is a basis for the complex vector space spanned by the set $\{f_1,f_2,...,f_l\}$.

Proof of (ii): The proof is by Aczél's method (see [1], [2], pp. 160-165) which uses induction and linear independence.

The proof is by induction on $s$. First we shall prove that the assertion is true for $s=1$. There exists a linearly independent function among $f_1,f_2,...,f_l$. We denote it by $f_{n_1}$; so we have $f_{n_1}(z)\not\equiv 0$ on $D$. Hence, by the continuity of $f_{n_1}$ on $D$, there exists a nonempty subdomain $D_1$ of $D$ such that $f_{n_1}(z)\neq 0$ on $D_1$. Hence, by (3), we have

$$\sum_{i=1}^{l}\left|\frac{f_i(z)}{f_{n_1}(z)}\right|^2 = \sum_{j=1}^{m}\left|\frac{g_j(z)}{f_{n_1}(z)}\right|^2 \qquad (9)$$

on $D_1$. Since $f_{n_1}$ is linearly independent on $D$ (i.e. $f_{n_1}(z)\not\equiv 0$) and since the dimension of the complex vector space spanned by the set $\{f_1,f_2,...,f_l\}$ is 1, $f_{n_1}$ is a basis for this complex vector space. So there exist complex coefficients $a_{p1}$ $(p\neq n_1)$ such that

$$f_p(z) = a_{p1}f_{n_1}(z) \text{ for all } p\neq n_1 \text{ and for all } z(moD). \qquad (10)$$

By (9), (10) we obtain on $D_1$

$$\sum_{j=1}^{m} \left| \frac{g_j(z)}{f_{n_1}(z)} \right|^2 = \sum_{p \neq n_1} |a_{p1}|^2 + 1. \tag{11}$$

Taking the Laplacians $\Delta = \dfrac{\partial^2}{\partial x^2} + \dfrac{\partial^2}{\partial y^2}$ $(z = x + iy)$ on both sides of (11) yields

$$\sum_{j=1}^{m} \left| \left( \frac{g_j(z)}{f_{n_1}(z)} \right)' \right|^2 = 0 \tag{12}$$

on $D_1$ since, by [5], p. 94, $\Delta |h(z)|^2 = 4|h'(z)|^2$, where $h$ is a regular function of $z$.

By (12) we have

$$\left( \frac{g_j(z)}{f_{n_1}(z)} \right)' = 0 \quad (j = 1,2,...,m)$$

on $D_1$ and so

$$g_j(z) = c_{j1} f_{n_1}(z) \quad (j = 1,2,...,m), \tag{13}$$

where, for each $j$ $(j = 1,2,...,m)$, $c_{j1}$ is a complex constant. By the Identity Theorem, (13) holds on the entire complex plane. Substituting (13) into (11) yields

$$\sum_{p \neq n_1} |a_{p1}|^2 + 1 = \sum_{j=1}^{m} |c_{j1}|^2. \tag{14}$$

By (13), (14) the assertion is true for $s = 1$.

We shall now prove that the assertion is true for $s = t+1$ if it is true for $s = t$ $(t = 1,2,...,l-1)$. By hypothesis we have on $D$

$$\sum_{i=1}^{t+1} |f_i(z)|^2 = \sum_{j=1}^{m} |g_j(z)|^2, \tag{15}$$

where $t+1 \leq m$. Since, by hypothesis, $f_{n_1}, f_{n_2}, ..., f_{n_t}, f_{n_{t+1}}$ are linearly independent on $D$, we have $f_{n_{t+1}} \not\equiv 0$ on $D$. Hence, by the continuity of $f_{n_{t+1}}$ on $D$, there exists a nonempty subdomain $D_{t+1}$ of $D$ such that $f_{n_{t+1}} \neq 0$ on $D_{t+1}$. So, by (15) we obtain

$$\sum_{\substack{1 \leq i \leq t \\ i \neq n_{t+1}}} \left| \frac{f_i(z)}{f_{n_{t+1}}(z)} \right|^2 + 1 = \sum_{j=1}^{m} \left| \frac{g_j(z)}{f_{n_{t+1}}(z)} \right|^2 \tag{16}$$

on $D_{t+1}$. Taking the Laplacians $\Delta = \dfrac{\partial^2}{\partial x^2} + \dfrac{\partial^2}{\partial y^2}$ of both sides of (16) yields on $D_{t+1}$

$$\sum_{\substack{1 \leq i \leq l \\ i \neq n_{t+1}}} \left| \left( \frac{f_i(z)}{f_{n_{t+1}}(z)} \right)' \right|^2 = \sum_{j=1}^{m} \left| \left( \frac{g_j(z)}{f_{n_{t+1}}(z)} \right)' \right|^2. \tag{17}$$

Since $f_{n_{t+1}}(z) \neq 0$ on $D_{t+1}$,

$$\left( \frac{f_i(z)}{f_{n_{t+1}}(z)} \right)' \quad \text{and} \quad \left( \frac{g_j(z)}{f_{n_{t+1}}(z)} \right)'$$

are regular on $D_{t+1}$ for each $i$ ($1 \leq i \leq l$, $i \neq n_{t+1}$) and $j$ ($j=1,2,...,m$).

We shall prove that the dimension of the complex vector space spanned by the set

$$\left\{ \left( \frac{f_i}{f_{n_{t+1}}} \right)' \ \middle| \ 1 \leq i \leq l, \ i \neq n_{t+1} \right\}$$

is $t$. First we shall prove that

$$\left( \frac{f_{n_i}}{f_{n_{t+1}}} \right)' \qquad (i=1,2,...,t)$$

are linearly independent on $D_{t+1}$, i.e.

$$\sum_{i=1}^{t} c_i \left( \frac{f_{n_i}(z)}{f_{n_{t+1}}(z)} \right)' = 0$$

on $D_{t+1}$ implies $c_i = 0$ ($i=1,2,...,t$). Here each of $c_i$ ($i=1,2,...,t$) is a complex constant.

By integration we have

$$\sum_{i=1}^{t+1} c_i f_{n_i}(z) = 0, \tag{18}$$

where $c_{t+1}$ is a complex constant. By the Identity Theorem, (18) holds on $D$.

Since $f_{n_1}(z), f_{n_2}(z), ..., f_{n_t}(z), f_{n_{t+1}}(z)$ are linearly independent on $D$, by (18) we obtain

$$c_i = 0 \quad (i=1,2,...,t) \quad (c_{t+1}=0).$$

Second we shall prove that each of

$$\left( \frac{f_p(z)}{f_{n_{t+1}}(z)} \right)' \quad (p \notin T),$$

where $T = \{n_1, n_2, ..., n_{t_1}\}$, can be represented by a linear combination of

$$\left( \frac{f_{n_i}(z)}{f_{n_{t+1}}(z)} \right)' \quad (i=1,2,...,t)$$

with complex coefficients on $D_{t+1}$. By Theorem (i) there exist complex coefficients $a_{pi}$ ($p \notin T$; $i=1,2,...,t+1$) such that

$$f_p(z) = \sum_{i=1}^{t+1} a_{pi} f_{n_i}(z) \quad (p \notin T) \tag{19}$$

holds on $D$. By (19) we obtain

$$\left( \frac{f_p(z)}{f_{n_{t+1}}(z)} \right)' = \sum_{i=1}^{t} a_{pi} \left( \frac{f_{n_i}(z)}{f_{n_{t+1}}(z)} \right)' \quad (p \notin T) \tag{20}$$

on $D_{t+1}$. By (20) each of

$$\left( \frac{f_p(z)}{f_{n_{t+1}}(z)} \right)' \quad (p \notin T)$$

can be represented by a linear combination of

$$\left( \frac{f_{n_i}(z)}{f_{n_{t+1}}(z)} \right)' \quad (i=1,2,...,t)$$

with complex coefficients on $D_{t+1}$. Consequently, the dimension of the complex vector space spanned by the set

$$\left\{ \left( \frac{f_i(z)}{f_{n_{t+1}}(z)} \right)' \;\middle|\; 1 \le i \le l, i \ne n_{t+1} \right\}$$

is $t$.  Hence, by (17), (20) and by the induction hypothesis there exists a constant complex $m \times t$ matrix $(c_{ji})$ satisfying

$$\sum_{p \notin T} |a_{pi}|^2 + 1 = \sum_{j=1}^{m} |c_{ji}|^2 \quad (i=1,2,...,t) \tag{21}$$

and

$$\sum_{p \notin T} a_{pi} \bar{a}_{pk} = \sum_{j=1}^{m} c_{ji} \bar{c}_{jk} \quad (i,k=1,2,...,t; \ i \neq k), \tag{22}$$

such that

$$\left( \frac{g_j(z)}{f_{n_{t+1}}(z)} \right)' = \sum_{i=1}^{t} c_{ji} \left( \frac{f_{n_i}(z)}{f_{n_{t+1}}(z)} \right)' \quad (j=1,2,...,m) \tag{23}$$

holds on $D_{t+1}$.

Integrating both sides of (23) yields on $D_{t+1}$

$$g_j(z) = \sum_{i=1}^{t+1} c_{ji} f_{n_i}(z) \quad (j=1,2,...,m), \tag{24}$$

where each of $c_{jt+1}$ $(j=1,2,...,m)$ is a complex constant.  By the Identity Theorem (24) holds on $D$.  Substituting (19), (24) back into (15) yields on $D$

$$\sum_{i=1}^{t+1} |f_{n_i}(z)|^2 + \sum_{p \notin T} \left| \sum_{i=1}^{t+1} a_{pi} f_{n_i}(z) \right|^2 = \sum_{j=1}^{m} \left| \sum_{i=1}^{t+1} c_{ji} f_{n_i}(z) \right|^2 . \tag{25}$$

Since

$$\left| \sum_{i=1}^{n} A_i \right|^2 = \sum_{i=1}^{n} |A_i|^2 + 2\mathrm{Re} \left( \sum_{1 \leq i < k \leq n} A_i \bar{A}_k \right),$$

by (25) we have

$$\sum_{i=1}^{t+1} |f_{n_i}(z)|^2 + \sum_{p \notin T}\left( \sum_{i=1}^{t+1} \left| a_{pi}f_{n_i}(z)\right|^2 \right.$$

$$+ 2\mathrm{Re}\left( \sum_{1 \le i < k \le t+1} a_{pi}f_{n_i}(z)\overline{a_{pk}f_{n_k}(z)}\right)\Bigg)$$

$$= \sum_{j=1}^{m}\left( \sum_{i=1}^{t+1}|c_{ji}f_{n_i}(z)|^2 + 2\mathrm{Re}\left( \sum_{1 \le i < k \le t+1} c_{ji}f_{n_i}(z)\overline{c_{jk}f_{n_k}(z)}\right)\right).$$

So we obtain on $D$

$$\sum_{i=1}^{t+1} |f_{n_i}(z)|^2 + \sum_{i=1}^{t}\left( |f_{n_i}(z)|^2 \sum_{p \notin T}|a_{pi}|^2 \right)$$

$$+ |f_{n_{t+1}}(z)|^2 \sum_{p \notin T}|a_{pt+1}|^2$$

$$+ 2\mathrm{Re}\left( \sum_{1 \le i < k \le t}\left( f_{n_i}(z)\overline{f_{n_k}(z)}\sum_{p \notin T}a_{pi}\bar{a}_{pk}\right)\right)$$

$$+ 2\mathrm{Re}\left( \sum_{i=1}^{t}\left( f_{n_i}(z)\overline{f_{n_{t+1}}(z)}\sum_{p \notin T}a_{pi}\bar{a}_{pt+1}\right)\right)$$

$$= \sum_{i=1}^{t}\left( |f_{n_i}(z)|^2 \sum_{j=1}^{m}|c_{ji}|^2 \right)$$

$$+ |f_{n_{t+1}}(z)|^2 \sum_{j=1}^{m}|c_{jt+1}|^2$$

$$+ 2\mathrm{Re}\left( \sum_{1 \le i < k \le t}\left( f_{n_i}(z)\overline{f_{n_k}(z)}\sum_{j=1}^{m}c_{ji}\bar{c}_{jk}\right)\right)$$

$$+ 2\mathrm{Re}\left( \sum_{i=1}^{t}\left( f_{n_i}(z)\overline{f_{n_{t+1}}(z)}\sum_{j=1}^{m}c_{ji}\overline{c_{jt+1}}\right)\right). \tag{26}$$

By (26), (21), (22) we have

$$\sum_{i=1}^{t+1} |f_{n_i}(z)|^2 + \sum_{i=1}^{t} \left( |f_{n_i}(z)|^2 \sum_{p \notin T} |a_{pi}|^2 \right)$$

$$+ |f_{n_{t+1}}(z)|^2 \sum_{p \notin T} |a_{pt+1}|^2$$

$$+ 2\mathrm{Re} \left( \sum_{1 \le i < k \le t} \left( f_{n_i}(z) \overline{f_{n_k}(z)} \sum_{p \notin T} a_{pi} \overline{a_{pk}} \right) \right)$$

$$+ 2\mathrm{Re} \left( \sum_{i=1}^{t} \left( f_{n_i}(z) \overline{f_{n_{t+1}}(z)} \sum_{p \notin T} a_{pi} \overline{a_{pt+1}} \right) \right)$$

$$= \sum_{i=1}^{t} \left( |f_{n_i}(z)|^2 \left( \sum_{p \notin T} |a_{pi}|^2 + 1 \right) \right)$$

$$+ |f_{n_{t+1}}(z)|^2 \sum_{j=1}^{m} |c_{jt+1}|^2$$

$$+ 2\mathrm{Re} \left( \sum_{1 \le i < k \le t} \left( f_{n_i}(z) \overline{f_{n_k}(z)} \sum_{p \notin T} a_{pi} \overline{a_{pk}} \right) \right)$$

$$+ 2\mathrm{Re} \left( \sum_{i=1}^{t} \left( f_{n_i}(z) \overline{f_{n_{t+1}}(z)} \sum_{j=1}^{m} c_{ji} \overline{c_{jt+1}} \right) \right),$$

and so

$$|f_{n_{t+1}}(z)|^2 + |f_{n_{t+1}}(z)|^2 \sum_{p \notin T} |a_{pt+1}|^2$$

$$+ 2\mathrm{Re} \left( \sum_{i=1}^{t} \left( f_{n_i}(z) \overline{f_{n_{t+1}}(z)} \sum_{p \notin T} a_{pi} \overline{a_{pt+1}} \right) \right)$$

$$= |f_{n_{t+1}}(z)|^2 \sum_{j=1}^{m} |c_{jt+1}|^2$$

$$+2\mathrm{Re}\left(\sum_{i=1}^{t}\left(f_{n_i}(z)\overline{f_{n_{t+1}}(z)}\sum_{j=1}^{m}c_{ji}\overline{c_{jt+1}}\right)\right).\tag{27}$$

Since

$$\overline{f_{n_{t+1}}(z)} = \frac{|f_{n_{t+1}}(z)|^2}{f_{n_{t+1}}(z)}$$

and

$$|f_{n_{t+1}}(z)|^2 \neq 0$$

on $D_{t+1}$, by (27) we have

$$\mathrm{Re}\left(\sum_{i=1}^{t}\left(\frac{f_{n_i}(z)}{f_{n_{t+1}}(z)}\left(\sum_{p \notin T}a_{pi}\overline{a_{pt+1}} - \sum_{j=1}^{m}c_{ji}\overline{c_{jt+1}}\right)\right)\right)$$

$$= \frac{1}{2}\left(\sum_{j=1}^{m}|c_{jt+1}|^2 - \sum_{p \notin T}|a_{pt+1}|^2 - 1\right)\tag{28}$$

on $D_{t+1}$.

The function

$$\sum_{i=1}^{t}\left(\frac{f_{n_i}(z)}{f_{n_{t+1}}(z)}\left(\sum_{p \notin T}a_{pi}\overline{a_{pt+1}} - \sum_{j=1}^{m}c_{ji}\overline{c_{jt+1}}\right)\right)$$

in (28) is regular on $D_{t+1}$. Hence, by a well-known theorem in analytic function theory and by (28), this function is merely a complex constant. So, by (28), we obtain on $D_{t+1}$

$$\sum_{i=1}^{t}\left(\left(\sum_{p \notin T}a_{pi}\overline{a_{pt+1}} - \sum_{j=1}^{m}c_{ji}\overline{c_{jt+1}}\right)f_{n_i}(z)\right) - Kf_{n_{t+1}}(z) = 0,\tag{29}$$

where $K$ is a complex constant with

$$\mathrm{Re}(K) = \frac{1}{2}\left(\sum_{j=1}^{m}|c_{jt+1}|^2 - \sum_{p \notin T}|a_{pt+1}|^2 - 1\right).\tag{30}$$

By the Identity Theorem (29) holds on $D$. Furthermore,

$f_{n_1}, f_{n_2}, ..., f_{n_t}, f_{n_{t+1}}$ are linearly independent on $D$.  Hence we have

$$\sum_{p \notin T} a_{pi} \overline{a_{pt+1}} - \sum_{j=1}^{m} c_{ji} \overline{c_{jt+1}} = 0 \quad (i=1,2,...,t), \quad K=0. \qquad (31)$$

By (21), (22), (31), (30) we have

$$\sum_{p \notin T} |a_{pi}|^2 + 1 = \sum_{j=1}^{m} |c_{ji}|^2 \quad (i=1,2,...,t+1)$$

and

$$\sum_{p \notin T} a_{pi} \overline{a_{pk}} = \sum_{j=1}^{m} c_{ji} \overline{c_{jk}} \quad (i,k=1,2,...,t+1; \ i \neq k).$$

Furthermore, by (24)

$$\begin{pmatrix} g_1(z) \\ g_2(z) \\ \cdot \\ \cdot \\ \cdot \\ g_m(z) \end{pmatrix} = \mathbf{C} \begin{pmatrix} f_{n_1}(z) \\ f_{n_2}(z) \\ \cdot \\ \cdot \\ \cdot \\ f_{n_{t+1}}(z) \end{pmatrix}$$

holds on $D$.  Here $\mathbf{C}=(c_{ji})$ $(j=1,2,...,m; \ i=1,2,...,t+1)$.  Thus the proof of the theorem is now completed.

## 3.  Proof of theorems A and B by using the theorem

Proof of Theorem B:  Since, by hypothesis, the set $\{f_1, f_2, ..., f_l\}$ is a basis for the complex vector space spanned by the set $\{f_1, f_2, ..., f_l\}$ (and so the dimension of this complex vector space is $l$), by a remark in section 1 and by (6), (7) we obtain

$$\sum_{j=1}^{m} c_{jp} \overline{c_{jq}} = \delta_{pq} \quad (p,q=1,2,...,l),$$

where $\delta_{pq}$ is the "Kronecker delta".  Q.E.D.

Theorem A follows from Theorem B.

## 4. A remark

The functional equation (3) is of Pexider type involving $l+m$ unknown functions $f_1, f_2, ..., f_l; g_1, g_2, ..., g_m$. By letting $s$ in the theorem range over the set $\{0,1,2,...,l\}$ the functional equation (3) is solved. This fact may be paraphrased as follows: the set of all systems of solutions of (3) is given by $\bigcup_{i=0}^{l} V_i$. Here we denote by $V_i$ the set of all systems of solutions of (3) when the dimension of the complex vector space spanned by the set $\{f_1, f_2, ..., f_l\}$ is $i$.

Department of Pure Mathematics
University of Waterloo
Waterloo, Ont., Canada
N2L 3G1

## References

[1] Aczél, J.: 1961, Sur une classe d'équations fonctionnelles bilinéaires à plusieurs fonctions inconnues. *Univ. Beograd. Publ. Elektrotehn. Fak., Ser. Mat. Fiz.* No. 61-64, 12-20.

[2] Aczél, J.: 1966, *Lectures on functional equations and their applications.* Academic Press, New York-London.

[3] Haruki, H.: 1982, A generalization of the Nevanlinna-Pólya theorem in analytic function theory. In: *Supplement to the Proc. of the 2nd World Conf. on Math. at the Service of Man,* Las Palmas, Spain.

[4] Nevanlinna, R. und G. Pólya: 1931, Unitäre Transformationen analytischer Funktionen. *Jber. Deutsch. Math.-Verein.* *40,* 80 (Aufgabe 103).

[5] Pólya, G. und G. Szegö: 1954, *Aufgaben und Lehrsätze aus der Analysis, I.* Springer-Verlag, Berlin.

[6]   Schmidt, H.: 1934, Lösung der Aufgabe 103, *Jber. Deutsch. Math.-Verein.* 43, 6-7.

L. Reich and J. Schwaiger

## On polynomials in additive and multiplicative functions

## 1. Introduction

In the theory of linear differential systems with constant coefficients ([2], 3.14), in the investigation of the partial differential equation of Peschl-Bauer, ([5], § 3), and in the theory of analytic iterations in rings of formal power series, theorems of the following type are useful:

(A) Let $\lambda_1,...,\lambda_n$ be $n$ distinct complex numbers and let $P(X, Y_1,..., Y_n)$ be a polynomial in $X$ and $Y_1,..., Y_n$. Suppose that $P(t, e^{\lambda_1 t},..., e^{\lambda_n t}) = 0$ for all $t \in \mathbb{C}$. Then we have $P = 0$.

(B) Let $\lambda_1,...,\lambda_n$ be complex numbers such that for all $(\alpha_1,...,\alpha_n) \in \mathbb{N}_0$ with $(\alpha_1,...,\alpha_n) \neq (0,...,0)$, $\sum_{i=1} \alpha_i \lambda_i = 0$ always holds. Moreover, let $P(X, Y_1,.., Y_n)$ be a polynomial with complex coefficients which vanishes if we substitute $X = t, Y_1 = e^{\lambda_1 t}$, $t \in \mathbb{C}$. Then we have again $P = 0$.

Theorems of the type (A) are needed, e.g., for the construction of a fundamental system of a linear differential system with constant coefficients.

In the usual proofs of such theorems differentiability or analyticity of the functions $t \to ct$, $t \to e^{\lambda_1 t}$ is used. But it turns out that we really need only the fact that these functions are solutions of Cauchy's famous functional equations, specifically, that $t \to ct$ is a solution of

$$f(t+s) = f(t) + f(s), \quad (t,s) \in \mathbb{C}^2,$$

(i.e. that $t \to ct$ is an *additive* function) and that $t \to e^{\lambda, t}$ is a solution of

J. Aczél (ed.), Functional Equations: History, Applications and Theory, 127-160.
© 1984 by D. Reidel Publishing Company.

$$g(t+s) = g(t)g(s), \quad (t,s)\in \mathbb{C}^2.$$

We shall call multiplicative functions the solutions of this Cauchy equation.

It is therefore our aim to prove (A) and (B) in such a manner that only the defining functional equations of additive or multiplicative functions are relevant, not any order of regularity of the functions considered, e.g. the analyticity of $t \to ct$ and of $t \to e^{\lambda t}$.

In order to be able to formulate our main results we have to recall the basic definitions of linear and algebraic dependence of functions.

DEFINITION 1. The functions $f_1,...,f_n$ from $\mathbb{C}$ to $\mathbb{C}$ will be called *linearly independent* over $\mathbb{C}$, if a relation $\alpha_1 f_1(t)+...+\alpha_n f_n(t)=0$ with constant coefficients $\alpha_j \in \mathbb{C}$ for each complex number $t$ implies that $\alpha_1=...=\alpha_n=0$. This can also be expressed in the following way: $f_1,...,f_n$ considered as elements of the $\mathbb{C}$-linear space of functions from $\mathbb{C}$ to $\mathbb{C}$, are linearly independent.

DEFINITION 2. The function $f_1,...,f_n$ from $\mathbb{C}$ to $\mathbb{C}$ will be called *algebraically independent* over $\mathbb{C}$, if each polynomial $P(Y_1,...,Y_n)$ over $\mathbb{C}$ such that $P(f_1(t),...,f_n(t))=0$, for all $t \in \mathbb{C}$, vanishes identically. Otherwise we call $f_1,...,f_n$ *algebraically dependent* over $\mathbb{C}$.

The set of all additive functions from $\mathbb{C}$ to $\mathbb{C}$ is evidently a linear subspace of the space mentioned in Definition 1. As to Definition 2, we make the remark that the functions $f_1,...,f_n$ over $\mathbb{C}$, if the operations are defined pointwise, generally generate not a field, but only a $\mathbb{C}$-algebra. Nevertheless, we use the expression "algebraic dependent" for them too. Obviously, the relation

"$P(f_1(t),...,f_n(t))=0$ for each complex $t$" is equivalent to the relation "$P(f_1,...,f_n)=0$" in the $\mathbb{C}$-algebra we mentioned above.

Now, using these definitions, we will show in § 2 that linearly independent additive functions $f_1,...,f_n$ are also algebraically independent over $\mathbb{C}$. If the functions $f_1,...,f_n$ are algebraically dependent, then one may ask what is the general algebraic relation over $\mathbb{C}$ which they fulfill. We will characterize these relations $P(f_1,...,f_n)=0$ by showing that the left hand sides $P$ form a certain polynomial ideal for which we construct an ideal basis starting out from the set of linear relations between $f_1,...,f_n$.

In Section 3 we will show that $m$ distinct multiplicative functions, none of them zero, are linearly independent. From this fact we will conclude: if the multiplicative functions $f_1,...,f_n$ are *multiplicatively independent*, i.e. if a relation

$$f_1^{\gamma_1} \cdot \cdots \cdot f_n^{\gamma_n} = 1$$

where $(\gamma_1,...,\gamma_n)\in Z^n$, is only possible if $(\gamma_1,...,\gamma_n)=(0,...,0)$, then the functions are also algebraically independent over $\mathbb{C}$; and vice versa. If the functions are multiplicatively and algebraically dependent, then we may ask again how the general algebraic relation over $\mathbb{C}$ between them looks. The answer is as follows. If, for a polynomial $P(Y_1,...,Y_n)$, $P(f_1,...,f_n)=0$, then there exists a monomial $Y_1^{\omega_1}\cdots Y_n^{\omega_n}$ such that $Y_1^{\omega_1}\cdots Y_n^{\omega_n}P(Y_1,...,Y_n)$ belongs to a certain polynomial ideal, for which an ideal basis can be constructed from the set of relations which hold between the $f_1,...,f_n$ in the sense of the theory of abelian groups.

These results are then put together to prove the results which are generalizations of (A) and (B) of this introduction. So, for instance, we will show the following as a generalization of (A). Suppose that $\phi_1,...,\phi_N$ are $N$ distinct multiplicative functions, none of them zero, and suppose that $f_1,...,f_n$ are additive functions, linearly independent over $\mathbb{C}$. Then a relation

$$\sum_{l=1}^{N} P_l(f_1,...,f_n)\phi_l = 0$$

with polynomials $P_1,...,P_N$ over $\mathbb{C}$ implies $P_1=...=P_N=0$.

Concerning (B), we have the following result: A set of functions consisting of $n$ linearly independent additive functions $f_1,...,f_n$ and $m$ multiplicatively independent multiplicative functions $g_1,...,g_m$ is algebraically independent over $\mathbb{C}$. Moreover, we again characterize the general algebraic relation which holds between the set of functions $f_1,...,f_n$, $g_1,...,f_m$ when the additive functions $f_1,...,f_n$ are not necessarily linearly independent and the multiplicative functions are not necessarily multiplicatively independent.

It is easy to verify that the left hand sides $P(f_1,...,f_n)$ of the algebraic relations $P(f_1,...,f_n)=0$ between additive functions are generalized polynomials in the sense of Orlicz and Mazur (cf. [4]). Therefore, we present in Section 6 a further generalization of our result on relations of the form $\sum_{l=1}^{N} P_l \phi_l = 0$ to the case of generalized polynomials $P_1,...,P_N$. We will finish this paper by a rather simple and genuine characterization, in the space of all generalized polynomials, of the functions $P(f_1,...,f_n)$, where $P$ is an ordinary polynomial over $\mathbb{C}$ and $f_1,...,f_n$ are additive functions using the so-called shift operators.

## 2. Algebraic relations between additive functions

THEOREM 1. Let $g_1,...,g_m$ $(m \geq 1)$ be additive functions, linearly independent over $\mathbb{C}$. Then these functions are algebraically independent over $\mathbb{C}$.

Proof. Consider a basis $(\xi_\alpha)_{\alpha \in I}$ of the linear space $\mathbb{C}$ over $\mathbb{Q}$. Each $\xi \in \mathbb{C}$ has a unique representation $\xi = \sum_{\alpha \in I} u_\alpha \xi_\alpha$, where $u_\alpha \in \mathbb{Q}$, and at most a finite number of $u$'s are different from zero. Since each $g_j$ is $\mathbb{Q}$-linear, we have therefore

$$g_j(\xi) = \sum_{\alpha \in I} u_\alpha \rho_{\alpha,j}, \quad j=1,...,m \qquad (1)$$

where $\rho_{\alpha,j}=g_j(\xi_\alpha)$, $\alpha\in I$. Among the vectors $(\rho_{\alpha,j})_{j=1,...,m}$, $\alpha\in I$, we choose a basis of the vector space they generate in $\mathbb{C}^m$; we denote this basis $\eta_{\alpha_1}=(\rho_{\alpha_1,j})_{j=1,...,m}\cdots\rho_{\alpha_r}=(\eta_{\alpha_r,j})_{j=1,...,m}$, when $r\geq m$. Each $\eta_\beta=(\rho_{\beta,j})_{j=1,...,m}$ where $\beta\in J:=I\backslash\{\alpha_1,...,\alpha_r\}$ has a unique representation

$$\eta_\beta = \sum_{l=1}^{r}\lambda_{\beta,l}\eta_{\alpha_l},$$

hence

$$\eta_\beta = \sum_{l=1}^{r}\lambda_{\beta,l}\rho_{\alpha_{l,j}} \quad (1\leq j\leq m),$$

where $\lambda_{\beta,l}\in\mathbb{C}$. We have consequently, in addition to (1),

$$g_j(\xi) = \sum_{l=1}^{r}(u_{\alpha_l}+\sum_{\beta\in J}\lambda_{\beta,l}u_\beta)\rho_{\alpha_{l,j}} \quad (l\leq j\leq m). \tag{2}$$

Using this notation we prove the following.

LEMMA 1. The additive functions $g_1,...,g_m$ are linearly independent over $\mathbb{C}$ if and only if $r=m$ and the linear forms

$$\Phi_1(X) = \sum_{l=1}^{r}\rho_{\alpha_{l,1}}X_l,.....,\Phi_m(X) = \sum_{l=1}^{r}\rho_{\alpha_{l,m}}X_l$$

in indeterminates $X_1,...,X_r$ over $\mathbb{C}$ are linearly independent.

Proof of Lemma 1. Suppose firstly that the linear forms defined in the lemma are linearly independent, and suppose that there exists a relation $\gamma_1\Phi_1(X)+...+\gamma_m\Phi_m(X)=0$, where $\gamma_{i_0}\neq 0$. Then, using the definition of the $\Phi_j$'s, we obtain

$$\sum_{j=1}^{m}\gamma_j\sum_{l=1}^{r}\rho_{\alpha_{l,j}}X_l = \sum_{l=1}^{r}(\sum_{j=1}^{r}\rho_{\alpha_{l,j}}\gamma_j)X_l = 0,$$

which is the same as

$$\sum_{j=1}^{m} \rho_{\alpha_{l,j}} \gamma_j = 0$$

for $l=1,2,...,r$. From (2) we therefore derive

$$\sum_{j=1}^{m} \gamma_j q_j(\xi) = \sum_{j=1}^{m} \gamma_j \sum_{l=1}^{r} (u_{\alpha_l} + \sum_{\beta \in J} \lambda_{\beta,l} u_l) \rho_{\alpha_{l,j}}$$

$$= \sum_{l=1}^{r} (\sum_{j=1}^{m} \gamma_j \rho_{\alpha_{l,j}})(u_{\alpha_l} + \sum_{\beta \in J} \lambda_{\beta,l} u_\beta) = 0,$$

which is valid for all $\xi \in \mathbb{C}$.

Conversely, assume now that $\sum_{j=1}^{r} \gamma_j q_j(\xi)=0$ for all $\xi \in \mathbb{C}$, where at least one $\gamma_j \neq 0$. This means in particular that, if we choose $u_\beta=0$ for all $\beta \in J$ and take $u_{\alpha_l} \in \mathbb{Q}$ arbitrarily, the relations

$$\sum_{l=1}^{r} u_{\alpha_l} \sum_{j=1}^{m} \gamma_j \rho_{\alpha_{l,j}} = 0$$

hold. Hence, by [3], p. 122 for instance, we have, selecting independent indeterminates $Y_1,.., Y_r$ over $\mathbb{C}$:

$$\sum_{l=1}^{r} (\sum_{j=1}^{m} \gamma_j \rho_{\alpha_{l,j}}) Y_l = 0,$$

which is of course the same as

$$\sum_{j=1}^{m} \gamma_j \Phi_j(Y_1,..., Y_r) = \sum_{j=1}^{m} \gamma_j (\sum_{l=1}^{r} \rho_{\alpha_{l,j}} Y_l) = 0,$$

where one $\gamma_j \neq 0$.

If the functions $g_1,...,g_m$ or equivalently the linear forms $\Phi_1,...,\Phi_r$ are now linearly independent, then the matrix

$$(\rho_{\alpha_{l,j}})_{\substack{j=1,...,r \\ l=1,...,m}}$$

has rank $m$, and since $r \leq m$, $r=m$ follows.

We continue now the proof of Theorem 1. Let $P(Y_1,..., Y_m)$ be a complex polynomial, such that $P(g_1(\xi),...,g_m(\xi))=0$ for all $\xi \in \mathbb{C}$, and suppose that $g_1,...,g_m$ are linearly independent additive functions. According to (2) we can write

$$P\left(\sum_{l=1}^{m}(u_{\alpha_{k\beta}}+\sum_J\gamma_{\beta,l}u_\beta)\rho_{\alpha_{l,1}},...,\sum_{l=1}^{m}(u_{\alpha_l}+\sum_{\beta\in J}\gamma_{\beta,l}u_\beta)\rho_{\alpha_{l,m}}\right)=0,$$

for all possible values of $u_{\alpha_l}$, $u_\beta$, therefore this is valid when we set $u_\beta=0$ for all $\beta\in J$. Hence, we deduce also, for independent indeterminantes, $X_1,...,X_m$,

$$P\left(\sum_{l=1}^{m}\rho_{\alpha_{l,1}}X_1,...,\sum_{l=1}^{m}\rho_{\alpha_{l,m}}X_l\right)=0.$$

We have already observed that the matrix $(\rho_{\alpha_{l,j}})$ has rank $m$ and so the substitution

$$(X_1,...,X_m)\rightarrow(\Phi_1(X),...,\Phi_m(X))$$

is a nonsingular homogenous linear transformation of the indeterminates which means also that $P(X_1,...,X_r)=0$ identically. This finishes the proof of Theorem 1.


THEOREM 2.  Let $g_1,...,g_m$ $(m\geq1)$ be additive functions from $\mathbb{C}$ to $\mathbb{C}$; suppose the $g_1,...,g_k$ are linearly independent over $\mathbb{C}$, while for each $j>k$ $(j\leq m)$ there is a (then unique) representation

$$g_j=\sum_{l=1}^{k}\alpha_{jl}g_l,\quad j=k+1,...,m. \tag{3}$$

Then, if $P(Y_1,...,Y_m)$ is a polynomial over $\mathbb{C}$ such that $P(g_1,...,g_m)=0$, it belongs to the ideal in the polynomial ring generated by

$$Y_{k+1}-\sum_{l=1}^{k}\alpha_{k+1l}Y_l,...,Y_m-\sum_{l=1}^{k}\alpha_{ml}Y_l.$$

Conversely, each polynomial belonging to this ideal vanishes if $g_i$ is substituted for $y_i$, $(i 1,...,m)$.

Proof. We need a generalization of the Taylor expansion of a polynomial, namely in terms of monomials in

$$Y_{k+1}-\sum_{l=1}^{k}\alpha_{k+1l}Y_{l},\dots,Y_{m}-\sum_{l=1}^{k}\alpha_{ml}Y_{l}$$

instead of monomials $Y_1^{\nu_1},\dots,Y_m^{\nu_m}$. $P(Y_1,\dots,Y_m)$ can be written in the form

$$P(Y_1,\dots,Y_m)=\sum_{\nu}P_{\nu}(Y_1,\dots,Y_{m-1})Y_{m}^{\nu},$$

and therefore we have

$$P(Y_1,\dots,Y_m)=\sum_{\nu}P_{\nu}(Y_1,\dots,Y_{m-1})[(Y_{m}-\sum_{l=1}^{k}\alpha_{m_l})+\sum_{l=1}^{k}\alpha_{m_l}Y_{l}]^{\nu}$$

$$=\sum_{\nu}P_{\nu}^{*}(Y_1,\dots,Y_{m-1})(Y_{m}-\sum_{l}\alpha_{m_l}Y_{l})^{\nu},$$

where $P_{\nu}^{*}$ denotes a polynomial in $Y_1,\dots,Y_{m-1}$. In the next step we expand in a similar manner each $P_{\nu}^{*}$ according to powers of

$$Y_{m-1}-\sum_{l=1}^{k}\alpha_{m-1l}Y_{l},$$

and so on until we have reached $Y_{k+1}$. So we obtain an expansion

$$P(Y_1,\dots,Y_m)=$$

$$\sum_{\alpha}Q_{\alpha}(Y_1,\dots,Y_k)(Y_{k+1}-\sum_{l=1}^{k}\alpha_{k+1l}Y_{l})^{\alpha_1}\dots(Y_{m}-\sum_{l=1}^{k}\alpha_{mk}Y_{l})^{\alpha_{m-k}}, \qquad (4)$$

where the $Q_{\alpha}$ are polynomials. More precisely (4) can be split into two terms:

$$P(Y_1,\dots,Y_m)=Q_{0\dots0}(Y_1,\dots,Y_k)+\sum_{|\alpha|>0}Q_{\alpha}(Y_1,\dots,Y_k)\Phi^{\alpha}(Y), \qquad (5)$$

where

$$\Phi^{\alpha}(Y)=(Y_{k+1}-\sum_{l=1}^{k}\alpha_{k+1l}Y_{l})^{\alpha_1}\dots(Y_{m}-\sum_{l=1}^{k}\alpha_{ml}Y_{l})^{\alpha_{m-k}}.$$

The assumptions of Theorem 2 imply now that

$$P(g_1,\dots,g_m)=0,$$

and hence, by (5), that

$$Q_{0\ldots0}(g_1,\ldots,g_k) = 0.$$

Since $g_1,\ldots,g_k$ are linearly independent, $Q_{0\ldots0} = 0$ by Theorem 1, and Theorem 2 is proved.

## 3. Linear independence and algebraic independence of multiplicative functions

In all what follows we will apply the following.

THEOREM 3. Let $f_1,\ldots,f_n$ be $n$ distinct multiplicative functions, $f_j \neq 0$, for $j=1,\ldots,n$. Then $f_1,\ldots,f_n$ are linearly independent over $\mathbb{C}$.

Proof. We deduce this result from a remark on solutions of difference equations.

LEMMA 2. Let $\lambda_1,\ldots,\lambda_s$ be $s$ distinct complex numbers, and suppose $\lambda_j \neq 0$ for $i=1,\ldots,s$. Denote by $\Phi_i: \mathbb{C} \to \mathbb{C}$ a nontrivial solution of the difference equation

$$\Phi_i(t+t_0) = \lambda_i \Phi_i(t), \quad i=1,\ldots,s, \tag{6}$$

where $t_0 = 0$ is the same in each equation. Then the functions $\Phi_1,\ldots,\Phi_s$ are linearly independent over $\mathbb{C}$.

Proof. For $s=1$ the assertion of Lemma 2 is trivial. We will proceed by induction on the number $s$ of equations (6). We therefore assume the lemma to be true for solutions of $k$ difference equations where $1 \leq k \leq s-1$, and have to prove it when $k=s$. Suppose now that there exists a nontrivial linear relation

$$\alpha_1 \Phi_1 + \ldots + \alpha_s \Phi_s = 0. \tag{7}$$

Here we can also restrict ourselves to the case when all $\alpha_j$, $j=1,\ldots,s$, are different from zero, for otherwise we would be in the case of less than $s$ difference equations (6). But then (7) can be written as

$$\Phi_1 = \beta_2 \Phi_2 + \ldots + \beta_s \Phi_s, \tag{8}$$

and the coefficients $\beta_2,\ldots,\beta_s$ are uniquely determined since $\Phi_s,\ldots,\Phi_s$ are linearly independent according to the assumption of our induction. Substitute $t+t_0$ in (8) for $t$. Because of (6) we obtain

$$\Phi_1(t) = \lambda_1^{-1} \lambda_2 \beta_2 \Phi_2(t) + \ldots + \lambda_1^{-1} \lambda_s \beta_s \Phi_s(t)$$

for each $t \in \mathbb{C}$, and therefore

$$\lambda_1^{-1} \lambda_j \beta_j = \beta_j, \quad j=2,\ldots,s.$$

Since $\beta_j \neq 0$ we obtain $\lambda_1 = \lambda_j$, $j=2,\ldots,s$, which contradicts the assumptions of Lemma 2. Hence $\Phi_1,\ldots,\Phi_s$ are linearly independent.

Let us return to the proof of Theorem 3 now. We have assumed that there exists a $t_0 \in \mathbb{C}$ for which $f_1(t_0) \neq f_n(t_0)$. We divide the set $\{1,\ldots,n\}$ of indices into disjoint subsets so that in each class $\{i_1,\ldots,i_{r_1}\}$ exactly those indices are collected for which $f_{i_1}(t_0) = \ldots = f_{i_{r_1}}(t_0)$, whereas $f_k(t_0) \neq f_{i_1}(t_0)$ if $k \neq \{i_1,\ldots,i_{r_1}\}$. Now, $t_0$ was chosen in such a way that at least two different nonvoid classes exist. Rearranging the indices if necessary we obtain

$$f_1(t_0) = \ldots = f_{r_1}(t_0) = \lambda_1$$
$$f_{r_1+1}(t_0) = \ldots = f_{r_1+r_2}(t_0) = \lambda_2$$

$$\cdot$$
$$\cdot \tag{9}$$
$$\cdot$$

$$f_{r_1+\ldots+r_{s-1}+1}(t_0) = \ldots = f_n(t_0) = \lambda_s,$$

where $s \geq 2$, and $\lambda_i \neq \lambda_j$ for $i \neq j$. Let us consider a linear relation for the multiplicative functions $f_j$:

$$\alpha_1 f_1 + ... + \alpha_n f_n = 0. \tag{10}$$

Define $\Phi_j$ for $1 \leq j \leq s$ by

$$\Phi_j = \alpha_{r_1 + .. r_{j-1}+1} f_{r_1 + .. r_{j-1}+1} \cdots + \alpha_{r_1 + ... + r_j} f_{r_1 + ... + r_j}. \tag{11}$$

These functions are solutions of the difference equations (6) because of (9), and (10) can be simply written as

$$\Phi_1 + ... + \Phi_s = 0. \tag{12}$$

Suppose that $\Phi_{l_1}, ..., \Phi_{l_r}$ $(r \geq 1)$ are those of the functions $\Phi_j$ which are not identically zero. Since the assumptions of Lemma 2 are fulfilled by these functions we have a contradiction, therefore $\Phi_k = 0$, for $k = 1, ..., s$, specifically

$$\Phi_1 = \alpha_1 f_1 + ... + \alpha_{r_1} f_{r_1} = 0. \tag{13}$$

This is again a relation of the type (10), but since $s \geq 2$, we have $r_1 < n$, so the relation (13) is "shorter" than (10). We can apply this consideration to (13), and after finitely many steps we reach $\alpha_1 f_1(t) = 0$, for all $t \in \mathbb{C}$, i.e. $\alpha_1 = 0$. But in the same manner $\alpha_2 = ... = \alpha_s = 0$ can be proven. Therefore Theorem 3 is true.


We list now some properties of multiplicative functions with respect to the abelian group they generate and the relation of this group to algebraic dependence. All multiplicative functions under consideration are assumed to be not identically zero which is the same as nowhere zero.

(i)   Multiplicative functions then form an abelian group, multiplication being defined pointwise: $(f \cdot g)(t) = f(t) \cdot g(t)$, $t \in \mathbb{C}$.

(ii)  The multiplicative functions $f_1, ..., f_m$ are algebraically dependent over $\mathbb{C}$ if and only if the multiplicative functions

$$(f_1^{\alpha_1} \cdot ... \cdot f_m^{\alpha_m})_{(\alpha_1, ..., \alpha_m) \in \mathbb{N}_0^m}$$

are linearly dependent over $\mathbb{C}$.

(iii) Property (ii) together with Theorem 3 implies: the multiplicative functions $f_1,...,f_m$ are algebraically independent over $\mathbb{C}$ if and only if the functions

$$(f_1^{\alpha_1}\cdot...\cdot f_m^{\alpha_m})_{(\alpha_1,...,\alpha_m)\in\mathbb{N}_0^m}$$

are distinct.

(iv) Property (iii) may be expressed in the following way which is sometimes more convenient for our purposes:

$$f_1^{\alpha_1}\cdot...\cdot f_m^{\alpha_m} \neq f_1^{\beta_1}\cdot...\cdot f_m^{\beta_m}$$

for

$$(\alpha_1,...,\alpha_m) \neq (\beta_1,...,\beta_m)\in \mathbb{N}_0^m$$

(even $\in \mathbb{Z}^m$) if and only if there is no

$$(\gamma_1,...,\gamma_m)\in \mathbb{Z}^m,$$

for which $f_1^{\gamma_1}\cdot...\cdot f_m^{\gamma_m}=1$, i.e. if and only if the functions $f_1,...,f_m$ are independent in the sense of the theory of abelian groups ("multiplicatively independent").

The proofs of these remarks are quite obvious. When put together, they yield the following.

THEOREM 4. The multiplicative functions $f_1,...,f_m$ $(m\geq 1)$ are algebraically independent over $\mathbb{C}$ if and only if they are multiplicatively independent.

## 4. Algebraic relations between multiplicative functions

Here we allow the multiplicative functions to be multiplicatively dependent and will describe the most general algebraic relation which they fulfill. Let us denote by $G$ the abelian group they define according to property (i) of Section 3. It is well known from the theory of finitely generated abelian groups (cf. [3], § 10, Th. 7) that among the $f_1,...,f_m$ there exist $r(>0)$ independent generators, where $r$ is maximal. In our case we have always $r>0$, unless $G=\{1\}$. Let $f_1,...,f_r$ be this maximal set of independent functions. We shall indicate why $r>0$. If $r=0$, this means that for each $f_k$ there exists an integer $m_k>0$ such that $f_k^{m_k}=1$. Each value taken by the function $f_k$ would be a root of unity of order $m_k$, and therefore the range of $f_k$ would be a finite set, which is impossible, if $f_k \neq 1$, according to [7], Hilfssatz 5 (ii).

We will now assume that in the sequel $m=r+2$. We do this without loss of generality and in order to simplify the exposition of theorems and proofs. Under this assumption $f_{r+1}$ satisfies an equation of the form

$$f_{r+1}^{l_{r+1}} = f_1^{\alpha_1}...f_r^{\alpha_r} \tag{14}$$

where $l_{r+1}\neq0$, $\alpha_j \in \mathbf{Z}$. The exponents $l_{r+1}$ of $f_{r+1}$ appearing in relations (14) form a $\mathbf{Z}$-ideal, and hence a principal ideal $(d_{r+1})$, where $d_{r+1}>0$. Therefore there exists a relation

$$f_{r+1}^{d_{r+1}} = f_1^{\beta_1}...f_r^{\beta_r} \tag{14'}$$

and $d_{r+1}$ is the minimal exponent of $f_{r+1}$ in such an equation. The relation (14') is uniquely determined by this fact, for otherwise $f_1,...,f_r$ would be multiplicatively dependent. Similarly there are (nontrivial) relations of the form

$$f_{r+2}^{l_{r+2}} = f_1^{\alpha_1} ... f_r^{\alpha_r} \cdot f_{r+1}^{\alpha_{r+1}}, \tag{15}$$

where $l_{r+2}=0$. The set of exponents of $f_{r+2}$ in relations of type (15) form a $\mathbf{Z}$-ideal, hence a principal ideal $(d_{r+2})$, where $d_{r+2}>0$, and this implies the existence of a relation

$$f_{r+2}^{d_{r+2}} = f_1^{\alpha_1}...f_r^{\alpha_r} \cdot f_{r+1}^{\alpha_{r+1}} \tag{15'}$$

with $\alpha_j \in \mathbf{Z}$. We apply (14') and division in $\mathbf{Z}$ with remainder to get $\alpha_{r+1}=qd_{r+1}+\gamma_{r+1}<d_{r+1}$ and

$$f_{r+2}^{d_{r+2}} = f_1^{\tilde{\gamma}_1}...f_r^{\tilde{\gamma}_r} \cdot f_{r+1}^{\tilde{\gamma}_{r+1}}, \tag{15''}$$

where $d_{r+2}$ is minimal and $0<\gamma_{r+1}<d_{r+1}$. Again, (15'') is uniquely determined by the properties of the exponents. The definition of

the numbers $d_{r+1}, d_{r+2}$ and $\gamma_{r+1}$ is also used in the following Lemma, the proof of which may be left to the reader.

LEMMA 3. A relation

$$f_1^{\omega_1} \dots f_r^{\omega_r} f_{r+1}^{\omega_{r+1}} f_{r+2}^{\omega_{r+2}} = f_1^{\rho_1} \dots f_r^{\rho_r} f_{r+1}^{\rho_{r+1}} f_{r+2}^{\rho_{r+2}} \tag{16}$$

where $\omega_j, \rho_j \in \mathbb{Z}$, $0 \le \omega_{r+1} < d_{r+1}$, $0 \le \omega_{r+2}, \rho_{r+2}$, $\rho_{r+2} < d_{r+2}$ implies $\omega_j = \rho_j$ $(1 \le j \le r+2)$.

We need furthermore to rewrite the formula (14'), (15"). Among the exponents $\beta_1, \dots, \beta_r$ of (14') there may occur some negative ones, say $\beta_{i_1}, \dots, \beta_{i_k}$ $(k > 0)$, then we may write instead of (14'), the following:

$$f_{i_1}^{\beta_{i_1}} \dots f_{i_k}^{\beta_{i_k}} f_{r+1}^{\Delta_{r+1}} = f_{i_{k+1}}^{\beta_{i_{k+1}}} \dots f_{i_r}^{\beta_{i_r}}, \tag{17}$$

where all $\beta_j$ are nonnegative integers and $\{i_1, \dots, i_r\}$ is a permutation of $\{1, \dots, r\}$. Similarly we deduce from (15") a relation

$$f_{j_1}^{\gamma_{j_1}} \dots f_{j_i}^{\gamma_{j_i}} f_{r+2}^{\Delta_{r+2}} = f_{j_{i+1}}^{\gamma_{j_{i+1}}} \dots f_{j_r}^{\gamma_{j_r}} f_{r+1}^{\gamma_{r+1}} \tag{18}$$

where each $\gamma_k > 0$, $\{j_1, \dots, j_r\}$ is a permutation of $\{1, \dots, r\}$ and $0 \le \gamma_{r+1} < d_{r+1}$ holds. Let us call (17) and (18) the distinguished relations in the group $G$. In connection with (17) and (18) we introduce two polynomials,

$$\Psi_1(Y) := Y_{i_1}^{\beta_{i_1}} \dots Y_{i_k}^{\beta_{i_k}} Y_{r+1}^{\Delta_{r+1}} - Y_{i_{k+1}}^{\beta_{i_{k+1}}} \dots Y_{i_r}^{\beta_{i_r}} \tag{19}$$

and

$$\Psi_2(Y) := Y_{j_1}^{\gamma_{j_1}} \dots Y_{j_i}^{\gamma_{j_i}} Y_{r+2}^{\Delta_{r+2}} - Y_{j_{i+1}}^{\gamma_{j_{i+1}}} \dots Y_{j_r}^{\gamma_{j_r}} Y_{r+1}^{\gamma_{r+1}}. \tag{20}$$

Now we are able to formulate the main result of this section.

THEOREM 5. Let $f_1,...,f_m$ be multiplicative functions; $G$, the multiplicatively written abelian group they generate. Assume $\mathrm{rk}\, G=r$ and $m=r+2$. Then a polynomial $P(Y_1,...,Y_{r+2})$ has the property $P(f_1,...,f_{r+2})=0$ if and only if there exists a monomial $Y_1^{\omega_1}\cdot...\cdot Y_r^{\omega_r}$ such that $Y_1^{\omega_1}\cdot...\cdot Y_r^{\omega_r}\cdot P(Y_1,...,Y_{r+2})$ belongs to the ideal in $\mathbb{C}[Y_1,...,Y_{r+2}]$ which is generated by $\Psi_1(Y)$ and $\Psi_2(Y)$.

Proof. The proof is again based on a certain generalization of the Taylor expansion of a polynomial, which is contained in the following.

LEMMA 4. To each polynomial $P(Y_1,...,Y_r)$ there exists a monomial $Y_1^{\omega_1}\cdot...\cdot Y_r^{\omega_r}$ such that $Y_1^{\omega_1}\cdot...\cdot Y_r^{\omega_r}P(Y_1,...,Y_r)$ has an expansion of the form

$$Y_1^{\omega_1}...Y_r^{\omega_r}P(Y_1,...,Y_{r+2}) \tag{21}$$

$$= \sum_{(\gamma,\mu,\nu,\rho)} {}^{*}P_{\gamma\mu\nu\rho}(Y_1,...,Y_r)\Psi_1^\gamma(Y)\Psi_2^\mu(Y)Y_{r+1}^\nu Y_{r+2}^\rho \tag{21}$$

where $*$ means that summation is extended over all $(\gamma,\mu,\nu,\rho)\in\mathbb{N}_0^4$ for which $\nu<d_{r+1}$, $\rho<d_{r+2}$. Each $P_{\gamma\mu\nu\rho}(Y_1,...,Y_r)$ is a polynomial in $Y_1,...,Y_r$ only.

Proof. We arrange $P(Y_1,...,Y_r,Y_{r+1},Y_{r+2})$ as a polynomial in $Y_{r+2}$:

$$P(Y_1,...,Y_{r+2}) = \sum_{\alpha=0}^{N} P_\alpha(Y_1,...,Y_{r+1})Y_{r+2}^\alpha.$$

Division of each exponent $\alpha$ by $d_{r+2}$ yields

$$\alpha = q_\alpha d_{r+2}+r_\alpha, \quad 0\le r_\alpha<d_{r+2},$$

and with the aid of (20) we find

$$P(Y_1,...,Y_{r+1},Y_{r+2})$$

$$= \sum_{\alpha} P_\alpha(Y_1,...,Y_{r+1})(Y_{j_1}^{\gamma_{j_1}}... Y_{j_1}^{\gamma_{j_1}})^{-q_\alpha}.$$

$$(\Psi_2(Y) + Y_{j_{l+1}}^{\gamma_{j_{l+1}}}... Y_{r+1}^{\gamma_{r+1}})^{q_\alpha} Y_{r+2}^{r_\alpha},$$

therefore

$$P(Y_1,...,Y_{r+2}) = \sum_{\nu,\mu} Q_{\nu,\mu}(Y_1,...,Y_{r+1})\Psi_2^\nu(Y) Y_{r+2}^\mu,$$

where the exponents $\mu$ are smaller than $d_{r+2}$ and the $Q_{\nu,\mu}(Y_1,...,Y_{r+1})$ are polynomials in $Y_{r+1}$ and rational functions in $Y_1,...,Y_r$, with a common denominator which is a monomial in $Y_1,...,Y_r$. We then arrange each $Q_{\nu,\mu}(Y_1,...,Y_{r+1})$ in the form

$$Q_{\nu,\mu}(Y_1,...,Y_{r+1}) = \sum_{\alpha} Q_{\nu,\mu,\alpha}(Y_1,...,Y_r) Y_{r+1}^\alpha,$$

and, recalling definition (19), we obtain by similar calculations

$$Q_{\nu,\mu,\alpha}(Y_1,...,Y_{r+1}) = \sum_{\gamma,\rho} Q_{\nu,\mu,\gamma,\rho}(Y_1,...,Y_r)\Psi_1^\gamma(Y) Y_{r+1}^\rho;$$

the exponents are smaller than $d_{r+1}$, each $Q_{\nu,\mu,\gamma,\rho}$ is a rational function, and these functions have a certain monomial in $Y_1,...,Y_r$ as common denominator. By substituting these expressions for $Q_{\nu,\mu,\alpha}$ into $Q_{\nu,\mu}$ and also into the sum representation of $P$, and multiplying by the common denominator $Y_1^{\omega_1}\cdot...\cdot Y_r^{\omega_r}$ we obtain (21). (It can be shown that the expansion (21) of $P$ is unique.)

These preparations now make the proof of Theorem 5 very easy. Let $P(Y_1,...,Y_r,Y_{r+1},Y_{r+2})$ be a polynomial which vanishes if we substitute $f_1,...,f_m$. We use the expansion (21) for this $P$ and divide the sum into two parts:

$$Y_1^{\mu_1} \cdots Y_r^{\mu_r} P(Y_1, \ldots, Y_{r+2})$$

$$= \sum_{\substack{\lambda, \mu, \nu, \rho \\ \lambda + \mu > 0}} P_{\lambda \mu \nu \rho}(Y_1, \ldots, Y_r) \Psi_1^\lambda(Y) \Psi_2^\mu(Y) Y_{r+1}^\nu Y_{r+2}^\rho$$

$$+ \sum_{\nu, \rho} {}^* P_{00\nu\rho}(Y_1, \ldots, Y_r) Y_{r+1}^\nu Y_{r+2}^\rho.$$

By definition of $\Psi_1$ and $\Psi_2$ the equations $\Psi_1(f_1, \ldots, f_{r+1}) = 0$ and $\Psi_2(f_1, \ldots, f_{r+2}) = 0$ hold, and by assumption on $P$ we obtain

$$\sum_{\nu, \rho} {}^* P_{00\nu\rho}(f_1, \ldots, f_r) f_{r+1}^\nu f_{r+2}^\rho = 0. \tag{22}$$

From (22) we can deduce $P_{00\nu\rho}(Y) = 0$, because, if each $P_{00\nu\rho}(Y)$ is written as a linear combination of distinct monomials,

$$P_{00\nu\rho}(Y_1, \ldots, Y_r) = \sum_\alpha c_{\alpha\nu\rho} Y_1^{\alpha_1} \cdots Y_r^{\alpha_r},$$

equation (22) is equivalent to

$$\sum_{\nu, \rho}^* \sum_\alpha c_{\alpha\nu\rho} f_1^{\alpha_1} \cdots f_r^{\alpha_r} f_{r+1}^\nu f_{r+2}^\rho = 0.$$

According to Lemma 3 the multiplicative functions

$$f_1^{\alpha_1} \cdots f_r^{\alpha_r} f_{r+1}^\nu f_{r+2}^\rho,$$

belonging to different exponents $(\alpha_1, \ldots, \alpha_r, \nu, \rho)$, are different, since $0 \le \nu < d_{r+1}$, $0 \le \rho < d_{r+2}$. If we now apply Theorem 3, we see that these functions are linearly independent over $\mathbb{C}$, hence $c_{\alpha\nu\rho} = 0$ for each index vector $(\alpha_1, \ldots, \alpha_r, \nu, \rho)$, and $P_{00\nu\rho}(Y) = 0$ for each $(\nu, \rho)$. This finishes the proof of Theorem 5.

## 5. Algebraic relations between addition and multiplicative functions

We shall now take the results of the previous sections and apply them to a finite set of additive and multiplicative functions.

THEOREM 6. Let $\phi_1,...,\phi_n$ be $n$ distinct multiplicative functions, none of them zero. By $f_1,...,f_m$ we denote linearly independent additive functions, and we let $P_1,...,P_n$ be polynomials in indeterminates $Y_1,...,Y_m$ over $\mathbb{C}$. Then, if the relation

$$\sum_{l=1}^{n} P_l(f_1,...,f_n)\phi_l = 0 \tag{23}$$

holds, the polynomials $P_1,...,P_m$ vanish identically.

Proof. In the case $n=1$ we have a relation of the form $P_1(f_1,...,f_m)\phi_1=0$. Since $\phi_1(t)\neq0$ for all $t$, this implies $P_1(f_1,...,f_m)=0$ and hence $P_1=0$ by Theorem 1. There is another special case where Theorem 6 has already been proven. We will introduce the concept of the formal degree of a polynomial $Q(Y_1,...,Y_m)$, say. If $Q=0$, the formal degree of $Q$ is any nonnegative integer. If $Q\neq0$, the formal degree of $Q$ is any nonnegative integer not smaller than $deg\,Q$. Now assume, that $\sum_{l=1}^{n}$ (formal degree of $P_l$) $=0$ for a possible determination of the formal degrees. This is possible only if either $P_l=0$ or $P_l$ is a constant. If the latter case occurs at least once, then (23) is a relation of the form (7) for distinct multiplicative functions, where the coefficients $\alpha_j$ are different from zero. But this possibility is excluded by Theorem 3. In order to prove Theorem 6 we may now proceed by induction on the expression $K=n+\sum_{l=1}^{n}$ (formal degree of $P_l$). The cases $K=0,1$ are already settled. We will now assume, that Theorem 6 is true for $K\leq N-1$, and we will prove it for $K=N$ ($N\geq2$). Moreover, we may restrict our attention to the cases $n\geq2$ and $N-n=\sum_{l=1}^{n}$ (formal degree of $P_l$ >0), and that each $P_l\neq0$, $1\leq l\leq n$. For otherwise we would consider a relation of the form (23) but with a $K'=n'+\sum_{l=1}^{n}$ (formal degree of $P_l\leq N-1$), since $n\geq2$. There exists a $t_0\in\mathbb{C}$ such that $\phi_1(t_0)\neq\phi_n(t_0)$. Let us write $\phi_j(t_0)=\lambda_j$ ($1\leq j\leq n$), and $f_j(t_0)=\mu_j$, $1\leq j\leq n$. We replace $t$ by $t+t_0$ in (23), which yields

$$0 = \sum_{l=1}^{n} P_l(f_1(t+t_0),...,f_m(t+t_0))\phi_l(t+t_0)$$

$$= \sum \lambda_l P_l(f_1(t)+\mu_1,...,f_m(t)+\mu_m), \quad t \in \mathbb{C},$$

and, if we subtract this from

$$0 = \sum \lambda_1 P_1(f_1(t),...,f_m(t))\phi_l(t), \tag{23}$$

we find the relation

$$0 = \lambda_1(P_1(f_1+\mu_1,...,f_m+\mu_m)-P_1(f_1,...,f_m))\phi_1$$

$$+ \sum_{l=2}^{n}(\lambda_l P_l(f_1+\mu_1,...,f_m+\mu_m)-\lambda_1 P_l(f_1,...,\phi_n))\phi_l. \tag{24}$$

Relation (24) is again a relation of the type (23). If we use the abbreviation

$$P_l^*(Y_1,...,Y_n) = \lambda_l P_l(Y_1+\mu_1,...,Y_m+\mu_m)-\lambda_1 P_l(Y_1,...,Y_m),$$

we can rewrite (24) as

$$\sum_{l=1}^{n} P_l^*(f_1,...,f_m)\phi_l = 0. \tag{25}$$

We will show now by induction that (25) is a relation for which the number

$$n + \sum_{l=1}^{n} \text{ (formal degree of } P_l^*)$$

is less than $N$, for an admissible determination of the formal degrees, and hence

$$P_l^*(Y) = 0, \quad l = 1,...,n. \tag{26}$$

For $l=1$, it is true that the formal degree of $P_l^*(Y)$ can be chosen to be less than $deg\ P_1(Y)$. If we apply the linear operator

$$P(Y) \rightarrow \lambda_1(P(Y_1+\mu_1,...,Y_m+\mu_m)-P(Y_1,...,Y_m))$$

to a single monomial

$$Y_1^{\alpha_1} \cdot ... \cdot Y_m^{\alpha_m},$$

we obtain

$$\lambda_1(Y_1+\mu_1)^{\alpha_1}\cdots(Y_n+\mu_n)^{\alpha_n}-\lambda_1 Y_1^{\alpha_1},\ldots,Y_n^{\alpha_n}$$

$$=\lambda_1{\sum_{\beta}}^* d_\beta Y_1^{\beta_1}\cdots Y_n^{\beta_n},$$

where summation $^*$ refers only to monomials of degree $\beta_1+\ldots+\beta_n\leq\alpha_1+\ldots+\alpha_n-1$. From the linearity of this operation we get the assertion for $l=1$. For $l\geq 2$ we show similarly that the formal degree of $P_l^*$ can be taken to be the same as the formal degree of $P_l$. We consider here the linear operator which maps $P(Y)$ be $\lambda_l P(Y_1+\mu_1,\ldots,Y_m+\mu_m)-\lambda_1 P(Y_1,\ldots,Y_n)$, and observe that it acts on a single monomial $Y_1^{\alpha_1}\cdots Y_n^{\alpha_n}$ as follows:

$$\lambda_l(Y_1+\mu_1)^{\alpha_1}\cdots(Y_n+\mu_n)^{\alpha_n}-\lambda_1 Y_1^{\alpha_1}\cdots Y_n^{\alpha_n}$$

$$=(\lambda_l-\lambda_1)Y_1^{\alpha_1}\cdots Y_n^{\alpha_n}+\sum_{\beta^*}d_\beta Y_1^{\beta_1}\cdots Y_n^{\beta_n},$$

where, for certain $l\geq 2$, $\lambda_l-\lambda_1\neq 0$, and $\sum *$ means summation over $\beta=(\beta_1,\ldots,\beta_n)$ such that $\beta_1+\ldots+\beta_n\leq\alpha_1+\ldots+\alpha_n-1$. From (27) the assertion follows for any polynomial $P_l$, and (25) is indeed a relation of type (24) for which the number $K$ is less than $N$. Therefore, $P_l^*=0$ for each $l$.

Here we have to distinguish two cases again, namely $\lambda_l\neq\lambda_1$ (which occurs for $l=n$) and $\lambda_l=\lambda_1$. In the first case we have $P_l=0$. If $P_l$ were different from zero, it would contain a monomial

$$Y_1^{\alpha_1}\cdots Y_n^{\alpha_m}$$

of highest degree, with a nonzero coefficient $c_\alpha\neq 0$. But (27) shows then that the coefficient of this monomial in $P_l^*(Y)$ would be $(\lambda_l-\lambda_1)c_\alpha\neq 0$, whereas $P_l^*=0$ has already been derived. Therefore we have shown that there exist indices $l$ (for which $\lambda_l\neq\lambda_1$) such that $P_l=0$, contradicting our assumption that $P_l\neq 0$, and Theorem 6 is proved.

An immediate consequence of Theorem 6 is as follows.

THEOREM 7. Let $f_1,...,f_n$ be linearly independent additive functions, and let $g_1,...,g_m$ multiplicatively independent multiplicative functions. Then the functions $f_1,...,f_m,g_1,...,g_m$ are algebraically independent.

Proof. We have already proved in Section 2 that the multiplicative functions

$$(g_1^{\alpha_1}...g_m^{\alpha_m})_{(\alpha_1,...,\alpha_n)\in\mathbb{N}_0^m}$$

are distinct over $\mathbb{C}$. Hence, Theorem 7 is an immediate corollary of Theorem 6.

If we do not suppose that the additive functions $f_1,...,f_n$ are linearly independent, and the multiplicative functions $g_1,...,g_m$ are multiplicatively independent, we may ask how the most general algebraic relation between them can be described. This can be done as follows.

THEOREM 8. Let $f_1,...,f_n$ be additive functions of which $f_1,...,f_s$ are linearly independent over $\mathbb{C}$, such that, for each $j>s$, there holds a relation

$$f_j = \sum_{i=1}^{s} \alpha_{ji}f_i, \quad \alpha_{ji}\in \mathbb{C}.$$

Denote by $\Phi_j(X)$, for $j=s+1,...,n$, the polynomial $X_j-\sum_{j=1}^{s}\alpha_{ji}X_i$. Furthermore, let $g_1,...,g_m$ be multiplicative functions, for which we assume the notations and assumptions of Theorem 5. Then a polynomial

$$P(X_1,...,X_n,Y_1,...,Y_m)$$

has the property

$$P(f_1,...,f_n,g_1,...,g_m) = 0,$$

if and only if there exists a monomial $Y_1^{\omega_1}\cdot...\cdot Y_r^{\omega_r}$ such that $Y_1^{\omega_1}\cdot...\cdot Y_r^{\omega_r}P(X,Y)$ belongs to the ideal generated by $\Phi_{s+1}(X),...,\Phi_n(X),\Psi_1(Y),\Psi_2(Y)$.

## 6. Generalized polynomials and multiplicative functions

In this section we generalize Theorems 3 and 6 by replacing the expression $P(f_1,...,f_m)$ (in Theorem 6 for instance) by generalized polynomials (from $\mathbb{C}$ to $\mathbb{C}$) in the sense of Orlicz and Mazur.

We will start by recalling the definition and the principal properties of generalized polynomials (cf. [4]). Let $V$ and $W$ be linear spaces over $\mathbb{Q}$ and let $n$ be a positive integer: A mapping $u: V^n \to W$ is called $n$-additive, if $u$ is additive in each component. It is well known that each such $n$-additive mapping is $\mathbb{Q}$-linear. By a 0-additive mapping we understand a constant mapping; a 1-additive mapping $u: V \to W$ is a $\mathbb{Q}$-linear mapping. Now we are going to define homogeneous polynomials $p: V \to W$. Such a mapping is called a homogeneous polynomial of degree $n$ ($n$-homogeneous polynomial) if there exists an $n$-additive mapping $u: V^n \to W$ such that $p = u \circ \delta_n$, where $\delta_n$ is the diagonal mapping, $\delta_n: V \to V^n$, $\delta_n(x)=(x,...,x)$ for all $x \in V$. By definition, $p(x)=u(x,...,x)$. If $p$ is an $n$-homogeneous polynomial, $p: V \to W$, then we can always find a symmetric $n$-additive mapping $u: V^n \to W$, such that $p=u \circ \delta_n$. Symmetric means that for all $x_1,...,x_n \in V$ and for each permutation $\pi$ of $\{1,2,...,n\}$

$$u(x_{\pi(1)},...,x_{\pi(n)}) = u(x_1,...,x_n).$$

This symmetric $n$-additive mapping is uniquely determined by $p$ and we will therefore denote it by $p$. 0-homogeneous polynomials are constant, 1-homogeneous polynomials are the same as linear functions. We are now going to list some simple properties of homogeneous polynomials. For each $n$-homogeneous polynomial $p=n \circ \delta_n$ the mapping $p_{x_1},...,x_n$ defined by

$$x \in V \to p(x,...,x,x_{i+1},...,x_n)$$

is a homogeneous polynomial of degree $i$, where $x_{i+1},...,x_n \in V$ can be chosen arbitrarily. The binomial theorem holds in the form

$$p(x+y) = \sum_{i=0}^{n} \binom{n}{i} p(x^{(i)}, y^{(n-i)}),$$

where

$$p(x^{(i)}, y^{(n-i)}) = p(\underbrace{x, \ldots, x}_{i \ times}, \underbrace{y, \ldots, y}_{(n-i) \ times}).$$

A generalized polynomial can now be defined as a mapping $p: V \to W$ with a representation

$$p = \sum_{i=0}^{n} p_i, \quad (n \geq 0)$$

where $p_i$ is an $i$-homogeneous polynomial. It can be shown that this decomposition of a generalized polynomial as sum of homogeneous polynomials is unique, i.e., if we have

$$\sum_{i=0}^{n} q_i = \sum_{i=0}^{n} p_i$$

when $p_i$, $q_i$ are $i$-homogeneous polynomials, then $p_i = q_i$ for $0 \leq i \leq n$. Suppose that $p$ is a polynomial $p = \sum_{i=0}^{n} p_i \neq 0$, then $deg \ p := \max \{i | p_i \neq 0\}$ is well defined, and is called the degree of $p$.

In particular, if we take $V = W = \mathbb{C}$, considered as a linear space over $\mathbb{Q}$, then the generalized polynomials from $\mathbb{C}$ to $\mathbb{C}$ are defined. Throughout this paper we have already met very often such generalized polynomials from $\mathbb{C}$ to $\mathbb{C}$, namely, the mapping $t \in \mathbb{C} \to P(g_1(t),...,g_m(t)) = P(f_1,...,f_m)(t)$, where $P$ is an ordinary polynomial over $\mathbb{C}$ ("a polynomial in the sense of algebra") and $g_1,...,g_m$ are additive functions. This can be shown as follows. For additive functions $g_1,...,g_m$ from $\mathbb{C}$ to $\mathbb{C}$ the mapping

$$(t_1,...,t_m) \in \mathbb{C}^m \to g_1(t_1) \cdot ... \cdot g_m(t_m) \in \mathbb{C}$$

is obviously $m$-additive, hence $t \to g_1(t) \cdot ... \cdot g_m(t)$ is an $m$-homogeneous polynomial, and the sum of such expressions is an $m$-homogeneous polynomial, too.

There is a close connection between generalized polynomials and the calculus of finite differences. Here we mention only the following facts. Let $p$ be a generalized polynomial of degree $n$, and $s$ a complex number. Then $\Delta_s p$, defined by $\Delta_s p(t) = p(t+s) - p(t)$, is again a polynomial, either 0 or of a degree smaller than $n$. This can be seen by the following calculation. If

$$p = \sum_{i=0}^{n} p_i,$$

where $p_i$ is $i$-homogeneous, then we have

$$\Delta_s p(t) = \sum_{i=0}^{n} (p_i(t+s) - p_i(t)) = p_n(t^{(n-1)}, s) + \sum_{i=0}^{n-2} q_i(t)$$

by application of the binomial theorem for generalized polynomials. Here the $q_i$'s are polynomials homogeneous of degree $i$, and the mapping $t \to p_n(t^{(n-1)}, s)$ is an $(n-1)$-homogeneous polynomial. Therefore $\Delta_s p$ is a polynomial of degree at most $n-1$ or zero.

Now we have made all preparations to formulate and prove Theorem 9.

THEOREM 9.    Let $f_1, \ldots, f_n$ be $n$ distinct multiplicative functions, none zero. By $p_1, \ldots, p_n$ we denote generalized polynomials from $\mathbb{C}$ to $\mathbb{C}$, and let us suppose that

$$\sum_{i=1}^{n} p_i f_i = 0.$$

Then all polynomials $p_i$ vanish identically.

REMARK.    This Theorem 9 obviously contains Theorem 3. But Theorem 6 is also a consequence, if we observe that the functions $P_i(f_1, \ldots, f_n)$, where $f_1, \ldots, f_n$ are linearly independent additive functions and $P_i$ is an ordinary polynomial, are themselves generalized polynomials. Furthermore, we have to apply Theorem 1 once more.

In order to prove Theorem 9 we need one more lemma.

LEMMA 5. Let $p: \mathbb{C} \rightarrow \mathbb{C}$ be a generalized polynomial and $f$ a multiplicative function, different from 1. If $p(t+s)f(s)-p(t)=0$ for all complex numbers $t$ and $s$, then $p$ is zero.

Proof. First, $f=1$ means that there exists an $s_0 \in \mathbb{C}$ such that $f(s_0)=1$. If $f(s_0)=0$, $f=0$, and the assumptions lead to $p(t+s_0)f(s_0)=p(t)=0$ for all $t$, then $p=0$. If $f(s_0)\neq 0$, then the assumption yields

$$p(t+s_0) = ap(t) \tag{28}$$

with $a\neq(f(s_0))^{-1}$, and $a=1$. Suppose now $p\neq 0$. Then we have a decomposition

$$p = \sum_{i=0}^{n} p_i$$

where $n\geq 0$, $p_i$ is $i$-homogeneous and $p_n\neq 0$. Substituting this representation into (28) we get

$$p_n(t+s_0)+p_{n-1}(t+s_0)+\ldots+p_0(t+s_0)$$
$$= ap_{n-1}(t)+\ldots+ap_0(t). \tag{29}$$

The mapping $t\rightarrow p_i(t+s_0)$ is a generalized polynomial which is either zero or has the degree $i$. This is a consequence of the binomial expansion

$$p_i(t+s_0) = \sum_{j=0}^{i} \binom{i}{j} p_i\big(t^{(j)}, s_0^{(i-j)}\big).$$

This formula shows even more precisely that the homogeneous part of $p_i(t+s_0)$ of degree $i$ is $p_i(t)$. The right hand side of (29) has therefore a $p_n$ as homogeneous part of maximal degree $n$, whereas the left hand side has $1 \cdot p_n$ as homogeneous component of maximal degree $n$. Because $a \neq 1$, and $p_n \neq 0$ we have a $p_n \neq 1 \cdot p_n$, a contradiction, since the decomposition into homogeneous polynomials is unique.

Now we are able to prove Theorem 9 by induction over $n$, the number of distinct multiplicative functions appearing in the relation

$$\sum_{i=1}^{n} p_i f_i = 0.$$

In the case $n=1$ everything is clear, since $f_1 = 0$ implies $f_1(t) = 0$ for all $t \in \mathbb{C}$. Let us now assume that the theorem proved for the case of $n$ distinct multiplicative functions. We will prove it for the case of $n+1$ multiplicative functions $f_1, \ldots, f_{n+1}$, none of them zero. Denote by $p_1, \ldots, p_{n+1}$ $n+1$ generalized polynomials and let us assume that

$$\sum_{i=1}^{n+1} p_i f_i = 0. \tag{30}$$

In this relation we must suppose that all $p_i$ are different from zero, for otherwise we would be in the case of a relation of less than $n+1$ summands and then all the generalized polynomials would already be 0 by the assumption. Among these relations (30) there exists one for which $deg\ p_1$ is minimal, and we assume that we are now considering such a relation with minimal $deg\ p_1$,

$$\sum_{i=1}^{n+1} q_i f_i = 0$$

say. From this relation we deduce, for all complex numbers $s$ and $t$,

$$\sum_{i=1}^{n+1} q_i(t+s) f_i(t) f_i(s) = 0 \tag{31}$$

where the functional equation of the functions $f_i$ has been used. We multiply the relation

$$\sum_{i=1}^{n+1} q_i(t) f_i(t) = 0 \tag{32}$$

by $f_1(s)$ and subtract from (31) in order to get

$$(q_1(t+s)-q_1(t)f_1(s))f_1(t)$$
$$+\sum_{i=2}^{n+1}(q_i(t+s)f_i(s)-q_i(t)f_1(s))f_i(t) = 0. \tag{33}$$

This is a relation of type (30), with generalized polynomials $p_{i,s}$; $i=1,...,n+1$. But among the properties of generalized polynomials we have listed that the polynomial $t\to[q_1(t+s)-q_1(t)]f_1(s)$ in $t$ is either zero or has a degree which is smaller than $deg\ q_1$. Because of the minimality of $deg\ q_1$ only

$$p_{1,s}(t) = (q_1(t+s)-q_1(t))f_1(s) = 0$$

for all $t\in\mathbb{C}$ can occur, and (33) is a relation of the form

$$\sum_{i=2}^{n+1}p_{i,s}f_i = 0$$

for $n$ generalized polynomials $p_{i,s}$. Hence, by induction, we have

$$0 = p_{i,s}(t) = q_i(t+s)f_i(s)-q_i(t)f_1(s)$$

for all complex numbers $t$ and $s$, or equivalently

$$q_i(t+s)\tilde{f}_i(s)-q_i(t) = 0, \quad i=2,...,n+1,$$

for all complex numbers $t$, where $\tilde{f}_i=f_i\cdot f_1^{-1}$ is a multiplicative function which is not identically zero. So, by Lemma 5, we get $q_i=0$ for $2\le i\le n$. But the relation (32) shows then that $q_1=0$ too, and we have reached a contradiction. Therefore our Theorem 9 holds for $n+1$ distinct multiplicative functions, and is therefore proved.

## 7. Generalized polynomials and polynomials in additive functions

In Section 6 we have seen that every polynomial $P(g_1(t),...,g_n(t))$ in additive functions $g_1,...,g_n$ with coefficients in $\mathbb{C}$ is a generalized polynomial.

Now the question arises whether there are generalized polynomials which are not polynomials in additive functions. To answer this we want to present a theorem characterizing polynomials in additive functions. We need some preparations for stating that theorem in a general context. Let $M$ be a commutative group, the law of composition being written additively. Furthermore let $V$ be a $\mathbb{Q}$-vector space. Then, as in the preceding section, one can define a mapping $p: M \to V$ to be a (generalized) polynomial of degree less or equal to $n$ if $\Delta_h^{n+1} p = 0$ is satisfied for all $h \in M$. Details can be found in [8], for instance.

Call a mapping $q: M \to V$ $i$-additive if $q$ is additive in every component. Then we have the following result ([8]): $p$ is a polynomial of degree less or equal to $n$ if and only if we have $p = p_0 + p_1 + ... + p_n$, where $p_i(x) = q_i(x,...,x)$ and $q_i$ is $i$-additive and symmetric for all $i$ with $0 \le i \le n$. Also the $q_i$'s are uniquely determined by $p$. Furthermore it is clear that the $q_i$'s are $i$-linear if $M$ is a $\mathbb{Q}$-vector space too, and that the set $p_n$ of polynomials of degree less or equal to $n$ is a $\mathbb{Q}$-vector space.

Taking $V = K$, $K$ a field with characteristic zero, we can easily see that, for arbitrary additive functions $g_1,...,g_m: M \to K$, for arbitrary integers $k_1,...,k_n \ge 0$ and for arbitrary constants $a \in K$, the mapping $t \to a g_1(t)^{k_1} g_2(t)^{k_2} ... g_m(t)^{k_m}$ is a polynomial of degree less or equal to $k_1 + ... + k_m$. Consequently every polynomial $P(g_1(t),...,g_m(t))$ in additive functions $g_i: M \to K$ with coefficients in $K$ is a polynomial. Fixing $h \in M$ and $p \in P_n$, the mapping $S_h(p)$, defined by $S_h(p)(x) = p(x+h)$, is also a generalized polynomial in $P_n$. This can be seen by looking at the representation $p = p_0 + ... + p_n$ above. The proof of the following theorem is to be found in [9].

THEOREM 10. Let $A: M \to Gl_n(K)$ be given, where $K$ is an algebraically closed field of characteristic zero and where $Gl_n(K)$ is the set of all regular $n \times n$ matrices with entries in $K$. Let $A$ fulfill the Cauchy equation

$$A(t+s) = A(t).A(s)$$

for all $t, s \in M$.

Then the following holds: There exists a regular matrix $T \in Gl_n(K)$ such that for $B(t) := TA(t)T^{-1}$ we have

$$B(t) = B_1(t) \oplus \ldots \oplus B_r(t) := diag\ (B_1(t),\ldots,B_r(t)),$$

where the matrices $B_i$ are of the form $B_i(t) = \Pi_i(t).\exp C_i(t)$ with homomorphisms $\Pi_i\colon M \to K^* = K/\{0\}$ and matrices $C_i(t)$ nilpotent upper triangular matrices fulfilling

$$C_i(t+s) = C_i(t) + C_i(s),$$

$$C_i(t)C_i(s) = C_i(s)C_i(t)$$

for all $i$ and all $t,s \in M$.

COROLLARY. The entries $a_{ik}(t)$ of $A(t)$ are of the form

$$a_{ik}(t) = \sum_{j=1}^{r} P_j^{(i,k)}(t)\Pi_j(t),$$

where the $P_j^{(i,k)}$ are polynomials in those additive functions appearing as entries in the matrices $C_i(t)$. (Expressions as above will be called exponential polynomials.)

Proof. This is an immediate consequence of Theorem 10 if we take into account that $\exp C_i(t)$ has polynomials in additive functions as entries.

THEOREM 11. Let $V$ be a subspace of finite dimension of $K^M$. If $V$ is invariant under the shift operators $S_h$ for all $h \in M$, then $V$ contains only exponential polynomials.

Proof. We follow the method used in [10]. Let $f_1,...,f_n$ be a basis of $V$ and let $h$ be an arbitrary element of $M$. Because $S_h f_i$ is contained in $V$ we get

$$S_h f_i = \sum_{j=1}^{n} a_{ij}(h) f_j$$

with certain coefficients $a_{ij}(h) \in K$. For $h, k \in M$ we write

$$(S_h \cdot S_k)(f_i) = S_h(S_k f_i) = S_h(\sum_{l=1}^{n} a_{il}(k) f_l)$$

$$= \sum_{l=1}^{n} a_{il}(k) \sum_{j=1}^{n} a_{lj}(h) f_j = \sum_{j=1}^{n} (\sum_{l=1}^{n} a_{il}(k) a_{lj}(h)) f_j$$

and

$$S_{h+k}(f_i) = \sum_{j=1}^{n} a_{ij}(h+k) f_j.$$

Because $S_h S_k = S_{h+k}$ and because $f_1,...,f_n$ form a basis of $V$ we have

$$a_{ij}(k+h) = \sum_{l=1}^{n} a_{il}(k) a_{lj}(h)$$

for all $i,j$ and all $h, k \in M$.

This means that the matrix function $A(t) = (a_{ij}(t))$ satisfies the Cauchy equation appearing in the above theorem. Furthermore all $A(t)$ are regular because $E = A(0) = A(t-t) = A(t)A(-t)$.

Applying the corollary, we conclude that the $a_{ij}$'s are exponential polynomials. As

$$S_h f_i = \sum_{j=1}^{n} a_{ij}(h) f_j,$$

written explicitly, means

$$f_i(x+h) = \sum_{j=1}^{n} a_{ij}(h) f_j(x)$$

for all $x$ and $h$, putting $x=0$, we get

$$f_i(h) = \sum_{j=1}^{n} a_{ij}(h) f_j(0).$$

Consequently each $f_i$ is a linear combination of exponential polynomials and is thus an exponential polynomial itself. Thus $V$ has a basis consisting of exponential polynomials, which means that all functions in $V$ are of the same form.

For $f \in K^M$, denote by $S(f)$ the subspace of $K^M$ generated by the set $\{S_h(f) \in M\}$ of translations of $f$.

Then we have the following.

THEOREM 12. Let $K$ be an algebraically closed field of characteristic zero and let $M$ be an abelian group. Furthermore, let $p$ be a generalized polynomial from $M$ to $K$.

Then $p$ is a polynomial in additive functions if and only if $S(p)$ is of finite dimension.

Proof. (a) Suppose $p = P(g_1,...,g_m)$, where $P$ is a polynomial in $m$ variables with coefficients in $K$ and $g_1,...,g_m \colon M \to K$ are additive.

For $k = (k_1,...,k_m) \in \mathbb{N}_0^m$ consider $g^k := g_1^{k_1} \cdot ... \cdot g_m^{k_m}$ and the vector space $V$ generated by the mappings $g^k$, $|k| \leq l$, where we choose $l$ in such a way that

$$p = \sum_{|k| \leq l} c_k g^k \quad (c_k \in K).$$

We want to show that $S_h(g^k) \in V$ for all $h \in M$ and all $k$ with $|k| \leq m$. We have that $g^k(x+h)$ equals

$$\left(g_1(x+h)\right)^{k_1} \cdot \left(g_2(x+h)\right)^{k_2} \cdot ... \cdot \left(g_m(x+h)\right)^{k_m}.$$

This means, as one can easily see, that

$$g^k(x+h) = g^k(x) + \sum_{|k'| < l} d_{k'}(h) g^{k'}(x)$$

with certain $d_{k'}(h) \in K$ or $S_h g^k \in V$. Consequently for all $h$ the translation $S_h p = \sum_k c_k S_h g^k$ also lies in $V$. Thus $S(p)$ as a subspace of the finite dimensional space $V$ is of finite dimension.

(b)  Now let $S(p)$ be of finite dimension. Because of the very definition of this space, $S(p)$ is invariant under translation. Thus by the previous theorem $p$ is an exponential polynomial:

$$p = \sum p_i \pi_i.$$

Without loss of generality we may assume that 1 is one of the $\pi_i$'s, say $\pi_1 = 1$. Then we have

$$(p_1 - f) \cdot \pi_1 + \sum_{i=2} p_i \pi_i = 0.$$

Using the independence theorem of Section 6, which can easily be generalized to fit our more general situation, we get $p_1 - f = 0$, or $f = p_1$. So $f$ is a polynomial in additive functions because $p_1$ is.

REMARK.  The proof of the first part did not make use of the fact that $K$ is algebraically closed. Thus we have that, for arbitrary fields of $K$ of characteristic zero, for arbitrary abelian groups $M$ and for any polynomial $p: M \rightarrow K$ in additive functions, the vector space $S(p)$ is of finite dimension.

Finally we are able to answer the question stated at the beginning of this section.

Let $M$ be a $\mathbb{Q}$-vectorspace of infinite dimension and let $K$ be a field of characteristic zero; then there are polynomials $p: M \rightarrow K$ which are not polynomials in additive functions.

We consider the following example:

Let $A$ be a basis of $M$ and define $Q: M \times M \rightarrow K$ by

$$q(a,b) := \begin{cases} 1 & a = b, \\ 0 & a \neq b, \end{cases}$$

and

$$q(x,y) = q(\sum_{a\in A} r_a a, \sum_{b\in A} s_b b) = \sum_{a\in A} r_a s_a.$$

Then $q$ is bilinear and symmetric. So $p: M \to K$ defined by $p(x)=q(x,x)$ is a polynomial. We claim that $p$ is not a polynomial in additive functions. Otherwise, by the remark above, we would have that $S(p)$ is of finite dimension. This would imply the existence of a nontrivial relation

$$\sum_{i=1}^{r} \lambda_i S_{a_i} p = 0,$$

where $r \geq 1$, $\lambda_i \in K$, $a_i \in A$ and $\lambda_i \neq 0$ for all $i$. Evaluating that sum at $a \in A$ we get

$$0 = \sum_{i=1}^{r} \lambda_i p(a_i + a) = \sum_{i=1}^{r} \lambda_i q(a_i + a, a_i + a)$$

$$= \sum_{i=1}^{r} \lambda_i (2 + 2q(a_i, a)).$$

Doing this for $a=a_1, a_2, ..., a_r$ ($A$ is infinite) and dividing by 2 we are led to the system

$$\sum_{i\neq j} \lambda_i + 2\lambda_j = 0 \quad (1 \leq j \leq 1).$$

This means

$$\lambda_j = -\sum_{i=1}^{r} \lambda_i$$

and $\lambda_i \neq \lambda_j$ for all $i,j$. Substituting this into one of the equations we get $(r+1)\lambda_j = 0$ or $\lambda_j = 0$ for all $j$, a contradiction.

So we have an example of a polynomial which is not a polynomial in additive functions.

Mathematisches Institut
Universität Graz
Brandhofg. 18
A-8010 Graz
Austria

# References

[1] Aczél, J.: 1966, *Lectures on functional equations and their applications.* Academic Press, New York-London.

[2] Arnol'd, V.I.: 1979, *Gewöhnliche Differentialgleichungen.* Deutscher Verlag der Wissenschaften, Berlin.

[3] Lang, S.: 1971, *Algebra.* 4th ed., Addison-Wesley, Reading, Mass.

[4] Mazur, S. and W. Orlicz: 1935, Grundlegende Eigenschaften der polynomischen Operationen I. *Studia Math. 5,* 50-68.

[5] Reich, L: 1972, Über multiplikative und algebraisch verzweigte Lösungen der Differentialgleichung $(1+\epsilon z\bar{z})^2 w_{z\bar{z}}+\epsilon n(n+1)w=0$. *Ber. Ges. Math. Datenverarbeitung Bonn 57,* 13-28.

[6] Reich, L.: 1971, Über analytische Iteration linearer und kontrahierender biholomorpher Abbildungen. *Ber. Ges. Math. Datenverarbeitung Bonn, 42.*

[7] Reich, L.: 1981, Über die allgemeine Lösung der Translationsgleichung in Potenzreihenringen. *Ber. Math. Stat. Gesellschaft, Forschungszentrum Graz 159.*

[8] Djokovic, D.Z.: 1969, A representation theorem for $(X_1-1)(X_2-1)...(X_n-1)$ and its applications. *Ann. Polon. Math. 22,* 189-198.

[9] McKiernan, M.A.: 1977, The matrix equation $a(x \circ y) = a(x)+a(x)+a(y)+a(y)$. *Aequationes Math. 15,* 213-223.

[10] McKiernan, M.A.: 1977, Equations of the form $H(x \circ y) = \sum_i f_i(x)g_i(y)$. *Aequationes Math. 16,* 51-58.

Cl. Alsina

# Truncations of distribution functions

*Abstract.* We characterize the truncation of distribution functions by solving a functional equation of Jensen-type.

Let $\Delta^+$ denote the space of all probability distribution functions of nonnegative random variables, i.e.,

$$\Delta^+ = \{F | F:[-\infty,\infty] \to [0,1], \; F(0)=0, \; F \text{ is nondecreasing}$$
$$\text{and left-continuous on } [-\infty,\infty)\}.$$

Among the elements of $\Delta^+$ are functions $\epsilon_a$ defined, for $a \geq 0$, by

$$\epsilon_a(t) = \begin{cases} 0, & t \leq a, \\ 1, & t > a. \end{cases}$$

We will denote by $\epsilon_\infty$ the null function in $\Delta^+$. We consider $\Delta^+$ endowed with the topology of weak-convergence and we remark that $\Delta^+$ is a lattice with respect to the usual pointwise operations

$$(F \vee G)(x) = Max(F(x), G(x)), \quad (F \curlywedge G)(x) = Min(F(x), G(x)).$$

DEFINITION 1. The *truncation* of distribution functions is the function $t: \overline{\mathbb{R}}^+ \times \Delta^+ \to \Delta^+$, defined by $t(a,F) = F \vee \epsilon_a$ for all $a \geq 0$ and $F \in \Delta^+$.

We summarize in the following theorem some properties of $t$.

J. Aczél (ed.), Functional Equations: History, Applications and Theory, 161-165.

THEOREM 1. The truncation function $t$ satisfies the following conditions:

1)  $t$ is continuous;

2)  $t(0,F)=\epsilon_0\geq t(a,F)\geq t(\infty,F)=F$;

3)  $t(a,F)\leq t(b,F)$ if $a\geq b$ and $t(a,F)\leq t(a,G)$ if $F\leq G$;

4)  $t(a,\epsilon_b)=\epsilon_{Min(a,b)}$;

5)  $t(b,t(a,F))=t(Min(a,b),F)$;

6)  $t(a,F\vee G)=t(a,F)\vee t(a,G)$;

7)  $t(a,F\times G)=t(a,F)\times t(a,G)$;

8)  $t(a,\dfrac{F+G}{2})=(t(a,F)+t(a,G))/2.$


Our chief concern in this paper is to prove that the conditions 1), 4) and 8) characterize the truncation function.


THEOREM 2. Let $T$ be a function from $\overline{\mathbb{R}}^+\times\Delta^+$, satisfying the following conditions:

a)  $T$ is continuous;

b)  $T(a,\epsilon_b)=\epsilon_{Min(a,b)}$, for all $a,b\geq 0$;

c)  $T(a,\dfrac{F+G}{2})=\dfrac{T(a,F)+T(a,G)}{2}$ for all $a\geq 0$ and $F,G\in\Delta^+$.


Then $T$ is the truncation function defined by $T(a,F)=F\vee\epsilon_a$, for all $a\geq 0$ and $F\in\Delta^+$.


Proof. If $T:\overline{\mathbb{R}}^+\times\Delta^+\to\Delta^+$ satisfies a), b) and c) then we have, by a) and b),

$$T(a,\epsilon_\infty) = w\text{--}\lim_{b\to\infty} T(a,\epsilon_b) = w\text{--}\lim_{b\to\infty} \epsilon_{Min(a,b)} = \epsilon_a. \qquad (1)$$

The conditions c) and (1) yield

$$T(a, \tfrac{1}{2}F) = T(a, \tfrac{1}{2}(F + \epsilon_\infty)) = \tfrac{1}{2}(T(a,F) + T(a,\epsilon_\infty))$$

$$= \frac{T(a,F) + \epsilon_a}{2}, \tag{2}$$

and (2), together with c), imply that, for any pair $F_1$ and $F_2$ in $\Delta^+$,

$$T(a, \tfrac{1}{2}F_1 + \tfrac{1}{4}F_2) = T(a, \tfrac{1}{2}(F_1 + \tfrac{1}{2}F_2))$$

$$= \frac{T(a,F_1) + T(a, \tfrac{1}{2}F_2)}{2}$$

$$= \frac{T(a,F_1)}{2} + \frac{T(a,F_2)}{4} + \frac{\epsilon_a}{4}. \tag{3}$$

It follows by induction that, for any $F_1, F_2, ..., F_n$ in $\Delta^+$, we have

$$T(a, \sum_{i=1}^{n} \frac{1}{2^i}F_i) = \sum_{i=1}^{n} \frac{1}{2^i} T(a,F_i) + \frac{1}{2^n}\epsilon_a. \tag{4}$$

Thus, if $T$ is continuous and satisfies (4), we get

$$T(a, \sum_{i=1}^{\infty} \frac{1}{2^i}F_i) = \sum_{i=1}^{\infty} \frac{1}{2^i} T(a,F_i), \tag{6}$$

for any sequence $\{F_i\}$ in $\Delta^+$. Next, let $\lambda$ be any number in $[0,1]$ and consider

$$\lambda = \sum_{i=1}^{\infty} \lambda_i/2^i,$$

a binary expansion of $\lambda$ with $\lambda_i \in \{0,1\}$, for all $i$. Using (6), we have the following chain of equalities:

$$T(a, \lambda F + (1-\lambda)G) = T(a, \sum_{i=1}^{\infty} \frac{1}{2^i}(\lambda_i F + (1-\lambda_i)G))$$

$$= \sum_{i=1}^{\infty} \frac{1}{2^i} T(a, \lambda_i F + (1-\lambda_i)G)$$

$$= \sum_{i=1}^{\infty} \frac{1}{2^i}(\lambda_i T(a,F)+(1-\lambda_i) T(a,G))$$

$$= \lambda T(a,F)+(1-\lambda) T(a,G). \tag{7}$$

Moreover, (7) implies as a particular case

$$T(a,\lambda F) = T(a,\lambda F+(1-\lambda)\epsilon_\infty) = \lambda T(a,F)+(1-\lambda) T(a,\epsilon_\infty)$$

$$= \lambda T(a,F)+(1-\lambda)\epsilon_a. \tag{8}$$

In the case where we have $\lambda$ and $\mu$ in $[0,1]$ with $0<\lambda+\mu\leq 1$, (7) and (8) yield:

$$\lambda T(a,F)+\mu T(a,b) = (\lambda+\mu)[\frac{\lambda}{\lambda+\mu} T(a,F)$$

$$+ (1-\frac{\lambda}{\lambda+\mu}) T(a,G)]$$

$$= (\lambda+\mu) T(a,\frac{\lambda}{\lambda+\mu}F+\frac{\mu}{\lambda+\mu}G)$$

$$= T(a,\lambda F+\mu G)-(1-\lambda-\mu)\epsilon_a. \tag{9}$$

Given $\lambda_1,\lambda_2,...,\lambda_n$ in $[0,1]$, with

$$0< \sum_{i=1}^{n} \lambda_i \leq 1,$$

and $a_1,a_2,...,a_n$ in $\mathbf{R}^+$, we can deduce from (9) that, for $F = \sum_{i=1}^{n} \lambda_i \epsilon_{a_i}$,

$$T(a,F) = T(a, \sum_{i=1}^{n} \lambda_i \epsilon_{a_i}) = \sum_{i=1}^{n} \lambda_i T(a,\epsilon_{a_i})+(1-\sum_{i=1}^{n} \lambda_i)\epsilon_a$$

$$= \sum_{i=1}^{n} \lambda_i \epsilon_{Min(a,a_i)}-\sum_{i=1}^{n} \lambda_i \epsilon_a+\epsilon_a$$

$$= (\sum_{i=1}^{n} \lambda_i \epsilon_{a_i}) \vee \epsilon_a = F \vee \epsilon_a \tag{10}$$

holds. Thus $T(a,F)=F\vee\epsilon_a$ for all $a\geq0$ and $F$ in

$$E=\left\{\sum_{i=1}^{n}\lambda_i\epsilon_{a_i}\,|\,n\in N,\lambda_1,...,\lambda_n\in[0,1],\sum_{i=1}^{n}\lambda_i\leq1,a_2,...,a_n\in\mathbb{R}^+\right\}.$$

Using the continuity of $T$, it is possible to conclude $T(a,F)=F\vee\epsilon_a$ for all $a\geq0$ and $F$ in $\Delta^+$ because any function $F$ in $\Delta^+$ can be obtained as the weak limit of a sequence of distribution functions in $E$. The theorem is thus proved.

Dep. Mathemàtiques i Estadisïca
(E.T.S.A.B.) Univ. Politècnica de Barcelona
Diagonal 649, Barcelona 28,
Spain

## References

[1] Aczél, J.: 1966, *Functional equations and their applications*. Academic Press, New York-London.

[2] Schweizer, B. and A. Sklar: 1982, *Probabilistic metric spaces*, Elsevier-North Holland, New York.

B.R. Ebanks

## Kurepa's functional equation
## on Gaussian semigroups

This is another in a series of papers (Kurepa [1956], Tamura [1957], J. Erdös [1959], Hosszú [1963], Aczél [1965], Jessen et al. [1968], Ebanks [1979, 1981, 1982, 1983]) on the functional equation

$$\Delta(a,b)+\Delta(ab,c) = \Delta(a,bc)+\Delta(b,c), \quad a,b,c \in S. \qquad (1)$$

Here $\Delta: S \times S \to G$, where $S$ is a Gaussian semigroup (defined below) and $G$ is a divisible abelian group (i.e. $nx=g$ can be solved for $x \in G$, given any $g \in G$ and any natural number $n$). Aside from its simplicity and symmetry, this equation is interesting because of the variety of contexts in which it has arisen. (Among them, in addition to those in the references above, are also characterizations of branching entropies in information theory.) One of these, the measurement of household utility, has been considered previously by the author [1981]. (See also Beckmann and Funke [1978].) There the variables (for example $a,b,c$ above) are "attractions" representing the attractiveness or utility of various products. These attractions are combined via some binary operation whenever two products are aggregated. If, for example, the operation is multiplication and the attractions are natural numbers or nonzero integers, then they form a Gaussian semigroup. Equation (1) with $S$ a Gaussian semigroup has been solved by the author in a previous paper [1982], but there the technique was rather complicated and not pretty. Here we give a much simpler proof of the result. It appears in Section 2, after a short section on Gaussian semigroups.

*J. Aczél (ed.), Functional Equations: History, Applications and Theory, 167-173.*

## 1. Gaussian semigroups

The following definitions and results come from Kurosh [1963], in which more can be found on Gaussian semigroups.

Let $S$ be an abelian semigroup with unit element 1 satisfying the cancellation law. A *divisor of unity* is an element $u$ for which $S$ contains an inverse element $u^{-1}$,

$$uu^{-1} = 1.$$

*The divisors of unity form a subgroup.*

If for $a, b \in S$ there is a $c \in S$ such that $a = bc$, then $b$ is a *divisor* of $a$. Two elements $a$ and $b$ are called *associates* if each is a divisor of the other,

$$a = bc, \qquad b = ad.$$

Then $cd = 1$, and *both $c$ and $d$ are divisors of unity*. The relation of being associates is an equivalence relation, so that $S$ splits into *classes of associate elements;* one of these is the group of divisors of unity. The relation 'divides' *partially orders* the set of classes of associate elements of $S$ if, for two such classes $A, B$, we put $A \leq B$ whenever at least one (and therefore every) element of $A$ divides at least one (and therefore every) element of $B$. The class of divisors of unity is the unique minimal element of this set.

An element $p \in S$ that is not a divisor of unity is called *irreducible* if its only divisors, apart from divisors of unity, are its associates, that is, if it always follows from

$$p = ab$$

that one of the elements $a, b$ is a divisor of unity and the other an associate of $p$. *If $p$ is irreducible, then so are all its associates.* The classes of associate irreducible elements (if any exist) are precisely the minimal elements of the partially ordered set *(poset)* of classes of associate elements, other than the class of divisors of unity.

Two compositions of an element $a$ into the product of several elements,

$$a = b_1 b_2 \ldots b_k \quad \text{and} \quad a = c_1 c_2 \ldots c_l,$$

are called *associate decompositions* if $k = 1$ and if, after a suitable change in numbering of factors in one decomposition, the elements

$b_i$ and $c_i$ $(i=1,2,...,k)$ are associates.

An abelian semigroup $S$ with unit element satisfying the cancellation law is called a *Gaussian semigroup* if every element $a$ of $S$, which is not a divisor of unity, can be decomposed into the product of irreducible elements and if any two such decompositions of $a$ are associates.

In addition to the multiplicative semigroup of nonzero integers mentioned in the introduction, other examples of Gaussian semigroups are: (i) all abelian groups, and (ii) the multiplicative semigroup of nonzero polynomials in an indeterminate $x$ with coefficients from a commutative field.

## 2. Main Result

Equation (1) will be solved in conjunction with

$$\Delta(a,b) = \Delta(b,a) \qquad a,b \in S. \tag{2}$$

This entails no loss of generality, as one can then obtain the general solution of (1) by adding an arbitrary antisymmetric biadditive two-place function to the general solution of the system (1), (2). (See Aczél [1965] or Ebanks [1982].)

THEOREM. Let $S$ be a Gaussian semigroup, $G$ a divisible abelian group. A map $\Delta: S \times S \to G$ satisfies the system (1), (2) if and only if there exists a map $f: S \to G$ such that

$$\Delta(a,b) = f(a)+f(b)-f(ab), \quad a,b \in S. \tag{3}$$

Proof. It is easily verified that any $\Delta$ of the form (3) satisfies (1), (2). The proof of the converse will be accomplished in four steps.

*Step 1.* We show first that (3) holds for $a, b$ in $U$, the group of divisors of unity in $S$. Since $U$ is an abelian group, this was proved by Jessen et al. [1968].

*Step 2.* We define $f$ on $P$, the set of irreducibles (primes) in $S$, in such a way that (3) holds whenever all terms in the equation are defined. $P$ splits into equivalence classes of associate elements. From each class $<p>$ in $P$ we select (arbitrarily) a designated representative $p' \in <p>$ and define $f(p')$ arbitrarily. For all $p \in <p>$ there exists a unique $u \in U$ such that $p = p' u$. Extend the domain of $f$ to all of $<p>$ via

$$f(p) := f(p') + f(u) - \Delta(p', u), \quad p = p' u. \tag{4}$$

[*Note*: If $u = 1$ in (4), then it reads $f(p') = f(p') + f(1) - \Delta(p', 1)$. This is valid, since (1) with $b = 1$ gives $\Delta(a, 1) = \Delta(1, c) =$ constant, and that constant must be $f(1)$ since $\Delta(1, u) = f(1)$ by Step 1.]

*Step 3.* We extend the domain of $f$ to all of $S$ (in such a way that (3) holds) by induction on the poset of classes of associate elements of $S$. The condition that allows us to make this induction is the minimal condition, which states that every nonempty subset $N$ of the poset has at least one element that is minimal (in $N$). The poset of classes of associates in a Gaussian semigroup satisfies this condition.

The map $f$ is already defined on $U \cup P$, so we may assume that $f$ is defined for all proper divisors of $a$ in $S$ so that (3) holds. If $a$ is not irreducible, then

$$a = bc \tag{5}$$

where $b$ and $c$ are proper divisors of $a$ in $S$. (Thus $f(b)$ and $f(c)$ are defined.) Define

$$f(a) := f(b) + f(c) - \Delta(b, c). \tag{6}$$

We must show that $f(a)$ is well-defined. That is, supposing

$$a = de,$$

where $d$ and $e$ are proper divisors of $a$ in $S$, we must show that

$$f(b) + f(c) - \Delta(b, c) = f(d) + f(e) - \Delta(d, e). \tag{7}$$

To begin, $b, c, d,$ and $e$ have decompositions into products of irreducibles,

$$b = p_1 p_2 \cdots p_k, \quad c = p_{k+1} p_{k+2} \cdots p_m,$$

$$d = q_1 q_2 \cdots q_l, \quad e = q_{l+1} q_{l+2} \cdots q_n.$$

Since $bc = a = de$, $p_1 p_2 \cdots p_m$ and $q_1 q_2 \cdots q_n$ are associate decompositions of $a$. Thus we may write

$$b = rs, \quad c = tu; \quad d = rt, \quad e = su.$$

(Here $r$ is a greatest common divisor of $b$ and $d$, etc.) Now (7) is equivalent to

$$f(rs) + f(tu) - \Delta(rs, tu) = f(rt) + f(su) - \Delta(rt, su). \tag{7'}$$

Since $rs, tu, stu$, etc. (but not $rstu$) are all proper divisors of $a$, we have, by the induction hypothesis and (1),

$$\Delta(rs, tu) - f(rs) - f(tu)$$
$$= \Delta(rs, tu) - [f(r) + f(s) - \Delta(r, s)] - f(t) + f(u) - \Delta(t, u)$$
$$= \Delta(r, s) + \Delta(rs, tu) + \Delta(t, u) - f(r) - f(s) - f(t) - f(u)$$
$$= \Delta(r, stu) + \Delta(s, tu) + \Delta(t, u) - f(r) - f(s) - f(t) - f(u)$$
$$= \Delta(r, stu) + \Delta(s, t) + \Delta(st, u) - f(r) - f(s) - f(t) - f(u).$$

But the final expression is symmetric in $s$ and $t$ (by (2)), hence we have (7'). Therefore $f(a)$ is well-defined by (6) whenever (5) holds.

Thus (3) is established if neither $a$ nor $b$ is a divisor of unity. Also, in Steps 1 and 2, (3) was established in case one of the elements $a, b$ is a divisor of unity and the other is a divisor of unity or an irreducible. Step 4 covers the only remaining case, and completes the proof.

*Step 4.* We check that (3) holds in case one of the elements $a, b$ is a divisor of unity and the other is not in $U \cup P$. Let $b = u \in U$, and let $a$ have irreducible decomposition

$$a = p_1 p_2 \cdots p_m \quad (m \geq 2, \text{ since } a \notin U \cup P).$$

Now, by (1) and the first three steps,

$$\Delta(a, u) = \Delta(p_1 p_2 \cdots p_m, u)$$
$$= \Delta(p_1 \cdots p_{m-1}, p_m u) + \Delta(p_m, u) - \Delta(p_1 \cdots p_{m-1}, p_m)$$
$$= f(p_1 \cdots p_{m-1}) + f(p_m u) - f(p_1 p_2 \cdots p_m u) + f(p_m)$$
$$+ f(u) - f(p_m u) - f(p_1 \cdots p_{m-1}) - f(p_m)$$
$$+ f(p_1 p_2 \cdots p_m)$$
$$= f(a) + f(u) - f(au).$$

Equation (3) is now established for all $a, b \in S$.

Department of Mathematics
Texas Tech University
Lubbock, TX 79409
U.S.A.

# References

[1] Aczél, J.: 1965, The general solution of two functional equations by reduction to functions additive in two variables and with the aid of Hamel bases. *Glasnik. Mat.-Fiz. Astronom. 20*, 65-72.

[2] Beckmann, M.J. and U.H. Funke: 1978, Product attraction, advertising, and sales: Towards a utility model of market behavior. *Z. Operations Research 22*, 1-11.

[3] Ebanks, B.: 1979, Branching measures of information on strings. *Canadian Math. Bull. 22*, 433-448.

[4] Ebanks, B.: 1981, A characterization of separable utility functions. *Kybernetika 17*, 244-255.

[5] Ebanks, B.: 1982, Kurepa's functional equation on semigroups. *Stochastica 6*, 39-55.

[6] Ebanks, B.: 1983, Measures of vector information with the branching property. *Kybernetika 6*, to appear.

[7] Erdös, J.: 1959, A remark on the paper "On some functional equations" by S. Kurepa. *Glasnik Mat.-Fiz. Astronom. 14*, 3-5.

[8] Hosszú, M.: 1963, On a functional equation treated by S. Kurepa. *Glasnik Mat.-Fiz. Astronom. 18*, 59-60.

[9] Jessen, B., J. Karpf and A. Thorup: 1968, Some functional equations in groups and rings. *Math. Scand. 22,* 257-265.

[10] Kurepa, S.: 1956, On some functional equations. *Glasnik Mat.-Fiz. Astronom. 11,* 3-5.

[11] Kurosh, A.G.: 1963, *Lectures on general algebra.* Chelsea, N.Y., pp. 71-77.

[12] Tamura, T.: 1957, Commutative nonpotent archimedean semigroup with cancellation law, I. *J. Gakugei Tokushima Univ. 8,* 5-11.

J. Aczél

## Some recent results on information measures, a new generalization and some 'real life' interpretations of the 'old' and new measures

**1.** The great and vocal 'pure' mathematician G.H. Hardy wrote (quoted, e.g., in [18], p.120) "a science is said to be useful if its development tends to accentuate the existing inequalities in the distribution of wealth, or more directly promotes the destruction of human life".

While the second kind of applications is expressly excluded from this Conference, our first example is intended to counteract, in a way, the first kind of applications mentioned by Hardy, since it is connected to the endeavour to diminish, rather than increase, the inequalities in the distribution of wealth.

Let me start with a story which, to the best of my knowledge, is true and certainly in character. In Lysenko's successful campaign to 'rejuvenate' (in "newspeak"), that is (in "oldspeak"), destroy Biology in the Soviet Union, rather few scientists objected to his 'theory' and alleged praxis (to great personal danger to the objectors). Some of the objections were raised about his experimental methods, which were statistically unsound. Lysenko countered, true to his logic, by attacking Statistics and Probability Theory (of which then and there Statistics was considered to be a subfield). Again true to his methods, for such an attack he needed a quotation from Marx, Engels, Lenin or Stalin. He found it in "science is the enemy of chance". Had he been successful, the great harm he had done would have spilled over to Probability Theory (and maybe all of Mathematics). But it was allegedly Kolmogorov who countered by acknowledging that science is the enemy of chance so, he is said to have continued, how does one fight an enemy? By exploring his positions; getting to know him well in order to defeat him. That is what Probability Theory does.

*J. Aczél (ed.), Functional Equations: History, Applications and Theory, 175-189.*
© *1984 by D. Reidel Publishing Company.*

Similarly, in order to fight social and economic inequalities, one should explore them and measure them scientifically.

Theil (e.g. [30,31,32]), Cowell (e.g. [13,14,15]), Gehrig (e.g. [16]) and others (e.g. [11], [17], [28], see also the bibliography of the latter) have used measures of equality and inequality formed in analogy to entropies in Physics and Information Theory. The latter are measuring uncertainty about or information expected from a system of events (say, an experiment). The most important of them, the *Shannon entropy*, is given by the formula

$$H_n(p_1,...,p_n) = -\sum_{k=1}^{n} p_k \log p_k \tag{1}$$

where $p_k$ is the probability of the $k$-th event (outcome of an experiment, forecast, message, etc.). So

$$\sum_{k-1}^{n} p_k = 1 \tag{2}$$

and $p_k > 0$ or $p_k \geq 0$. In the latter case the convention

$$0 \log 0 := 0 \tag{3}$$

is adopted. The log in these formulas may be base 2 or base $e$ or base 8, or base 10, ... as long as one keeps always the same base in the same calculation. The maximum value of (1) under the condition (2) is $\log n$, so

$$I_n(p_1,...,p_n) = \log n - H_n(p_1,...,p_n)$$
$$= \log n + \sum_{k=1}^{n} p_k \log p_k \tag{4}$$

is always nonnegative.

Theil [30] uses this formula (4) as a measure of inequality. If, for instance, inequality of wealth is to be measured, possibly for individuals but mostly for entities (groups), one takes, in (4), $p_k = w_k/w$, where $w_k$ is the wealth of the $k$'th entity ($k=1,...,n$) and $w = \Sigma w_k$ the total wealth of the community. These $p_k$ certainly satisfy the above conditions, although Theil was criticized since these are not probabilities. On the one hand, this is a matter of interpretation (we will return to this point later), on the other it ignores that the above measure is supposed to be just analogous, not identical to the information theoretical entropy. In accordance with what one expects from a measure of inequality, the nonnegative quantity (4) is 0 if $p_1=p_2=...=p_n$ (there is no inequality in wealth between the $n$ entities if each has equal share in the wealth). Theil emphasizes the equally natural property of (4) which corresponds to the so called strong additivity of entropies (cf. [7]). Suppose that we have $mn$ entities which fall into $n$ groups, each containing $m$ entities. Then we can consider both between-group and within-group inequalities. Speaking again about wealth, if the $k$-th entity in the $j$-th group has the wealth $w_{jk}$ then $w_j = \sum_{k=1}^{m} w_{jk}$ is the wealth of the $j$-th group and, again, $w = \sum_{j=1}^{n} w_j$ the total wealth of the community. By writing $p_j=w_j/w$ (the wealth of the $j$-th group relative to the community) and $q_{jk} = w_{jk}/w_j$ (the wealth $w_{jk}$ relative to the wealth $w_j$ of the $j$-th group), we have $\sum_{j=1}^{n} p_j = 1 = \sum_{k=1}^{m} q_{jk}$ and, from (4),

$$I_{mn}(p_1 q_{11},...,p_1 q_{1m},.....,p_n q_{n1},...,p_n q_{nm})$$

$$= I_n(p_1,...,p_n) + \sum_{j-1}^{n} p_j I_m(q_{j1},...,q_{jm}).$$

The first term on the right hand side is the measure of the between-groups inequality of wealth, while the $I_m$'s in the last term measure the within-group inequality and the sum there is the average within-group inequality.

The following generalization of (5) would be equally natural, but is not satisfied by (4), though it is satisfied by (1) which is the 'measure of equality'. We have now $N = \sum_{j=1}^{n} m_j$ entities which fall into $n$ groups containing $m_1, m_2,...,m_n$ entities. Then, again, the total (this time) equality of wealth is the sum of the between-groups equality and of the average within-group equality:

$$H_N(p_1 q_{11},...,p_1 q_{1m_1},.....,p_n q_{n1},...,p_n q_{nm_n})$$

$$= H_n(p_1,...,p_n) + \sum_{j=1}^{n} p_j H_{m_j}(q_{j1},...,q_{jm_j})$$

$$(\sum_{j=1}^{n} p_j = 1 = \sum_{k=1}^{m_j} q_{jk}).$$

(There *does* exist a similar formula for the measure of inequality (4) too, but there the $m_1,...,m_n$ enter inconveniently also into the *first* term on the right hand side [30].)

The measure of equality (1), which Theil [31] preferred later, has also the useful property

$$H_{n+1}(p_1 q_1, p_1 q_2, p_2,...,p_n) = H_n(p_1, p_2,...,p_n)$$
$$+ p_1 H_2(q_1, q_2)$$
$$(p_1+...+p_n=1=q_1+q_2; \; n=2,3,...) \tag{7}$$

which is the special case $N=n+1$, $m_1=2$, $m_2=...=m_n=1$ of (6), if we write $q_k$ for $q_{1k}$ $(k=1,2)$ and agree in $H_1(1) = 0$ (just one entity which has all wealth: no equality – no inequality either, since $I_1(1) = \log 1 - H_1(1) = 0$). The equation (7) is called *recursivity* since it clearly can be used to construct all $H_n$ if $H_2$ is known. It describes the effect on the equality measure of splitting one entity (and its wealth) into two.

In older characterizations of the Shannon-entropy (1) the extensibility

$$H_{n+1}(p_1,...,p_n,0) = H_n(p_1,...,p_n)$$

was also used. This makes sense for the information theoretical entropy [with the definition (3)] but not for measures of equality (or inequality): even if entities with 0 wealth were considered, it would not be natural to assume that increasing the group by entities with 0 wealth would leave the equality (or inequality) of wealth unchanged.

Fortunately, recent results [22,24,25,6,9] show that, with the aid of *functional equations*, Shannon's (and other similar) entropies can be characterized without using expansibility and 0 probabilities at all, by the recursivity (7) and by symmetry (which just states that the measure is not dependent upon the order in which we have labelled the entities) with, say, continuity or boundedness added, for instance

$$0 \leq H_n(p_1,...,p_n) \leq \log n.$$

(That is, up to a nonnegative multiplicative constant, *only* (1) satisfies these conditions.)

Moreover, in [24,25,6,9] the same problem is solved for measures depending upon more than one probability distribution (any number). These, like the directed divergence $\Sigma p_k \log (p_k/q_k)$ (note that also (4) is of the form with $q_k = 1/n; \quad k=1,2,...,n$), or the information improvement (Theil's terminology [30]) $\Sigma p_k \log(q_k/r_k) \; \Sigma p_k = \Sigma q_k = \Sigma r_k = 1$) have also interesting applications in Economics (cf., e.g. [30,31]). Here the restriction to positive $p_k, q_k, r_k$ is even more important: these measures are not defined if $q_k = 0$ (or $r_k = 0$) but $p_k \neq 0$, and even the convention (3) cannot fix this.

Other, more unusual applications of information measures are in Ecology [26,27], Psychology [10] and even detective work in tracing down criminals (! [33]: the author actually was employed by the Home Office in London; also [12]).

Another application, in forecasting theory, is known under the catchy name 'how to keep the expert (or forecaster) honest'. There we have $n$ possible situations (weather, market development, possible outcomes of an experiment, etc., all more or less serving mankind). They have probabilities $p_1,...,p_n$ (for instance, in the sense of frequencies or in a sense to be discussed later), but we don't know them. So we ask an expert who gives us the (finite) probability distribution $q_1, q_2,..., q_n$ and agrees to be paid afterwards (!) the amount $f(q_k)$ if the $k$-th situation has been realized. Thus the expert's expected gain is

$$\sum_{k=1}^{n} p_k f(q_k).$$

We 'keep the expert honest' by maximizing this amount when $q_k = p_k$ that is, when we were told the truth. In other words, the 'payoff function' $f$ should be chosen so that it satisfies the *functional inequality*

$$\sum_{k=1}^{n} p_k f(q_k) \leq \sum_{k=1}^{n} p_k f(p_k)$$

$$(p_k > 0, q_k > 0, \sum_{k=1}^{n} p_k = \sum_{k=1}^{n} q_k = 1). \qquad (8)$$

While just ten years ago it was a fairly difficult (though known) theorem, it is now easy to prove (e.g. [1,7]) that, even *for a fixed* $n > 2$, the inequality (8) holds (if and) only if $f(q) = a\log q + b$, where $a \geq 0$, $b$ are constants. With this $f$, the right hand side of (8) goes over into

$$a \sum_{k=1}^{n} p_k \log p_k + b \quad (p_k > 0, \sum_m^n p_k = 1, \ n \text{ fixed}). \tag{9}$$

We notice immediately the connection of (9) with (1) and (4). (Since $a \geq 0$, the amount (9) will be nonnegative – and no expert worth his salt will accept a deal in which his expected earnings are negative – only if $b \geq a \log n$ and can also reach 0 if $b = a \log n$ as in (4).)

**2.** While many other 'applications' of information theory were and some are due to the "bandwagon effect" exposed by Shannon [29], using fashionable buzz-words like entropy, redundancy, etc., the above, in my opinion, do not fall into this category. Shannon emphasizes that "the hard core of information theory is, essentially, a branch of Mathematics, a strictly deductive system; a thorough understanding of the mathematical foundations and its communication application is surely a prerequisite to other appliations." We did not go here into the communications applications: they certainly are (mostly) at the service of mankind; they are also quite widely known by now. But in the other applications mentioned here, we have kept our perspective by emphasizing that some natural properties of the quantities in question (like symmetry, recursivity and continuity or boundedness of 'equality measures') determine them uniquely. So that is the form they *must* have and if it is the same or similar to the measures originally introduced in information theory for the purpose of communications applications, so be it: if something swims like a duck, eats like a duck, walks like a duck ... then it is a duck.

There is another criticism, mentioned before, that the $p_k$ are not really probabilities. Much of this is connected to the outdated or at least partisan view that probabilities are 'limits' of frequencies (in a certain sense, difficult to define since late large variations are always possible), so only such events which can be repeated infinitely often (or at least as often as we want to) can have probabilities. But this would be analogous to calling lengths, temperatures, etc. measurable only if they can be measured arbitrarily often.

Since Kolmogorov's fundamental work [21] in the 1930's, it is more or less accepted that probability is what satisfies certain conditions (axioms) and for our purposes only $\sum p_k = 1$ and $p_k > 0$ (or $p_k \geq 0$) are required. The above relative wealth quotients $p_k = w_k / \Sigma w_j$ certainly satisfy these and this could justify their use in this context. Anyway, whether they are probabilities or not, as explained before, the expressions involving them are useful simply because they, and under some circumstances only they, have certain desirable properties.

Some of the above argument came up also as explanation for the need for a 'theory of information without probability'. While the reasoning, that some events furnish information even though they have no probabilities because they cannot be repeated, may be answered as above, the main objection against that theory seems to be that it does not lead far enough (which, on the other hand, could be construed also as an incentive to work on it harder). We will not deal here further with that subject, but will rather give some details on another generalization of the probabilistic theory of information, the 'mixed theory'.

In the *mixed theory of information*, the measures of information are permitted to depend both on the events (messages, outcomes of an experiment, weather, market situations, etc.) themselves and on their probabilities (or similar parameters). For this we have to grasp mathematically the concept of an 'event'. It seems that they can be considered adequately as elements of a *ring of sets*, really of subsets of a comprehensive set (universe) $S$ which contains, with any two subsets, also their union ($\cup$) and difference ($\backslash$) therefore also their intersection ($\cap$) and the (empty) set $\emptyset$. If also $S$ belongs to the ring, then it is *Boolean*. So we are dealing now with (real valued) measures

$$H_n \begin{Bmatrix} E_1, E_2, \dots, E_n \\ p_1, p_2, \dots, p_n \end{Bmatrix} \quad (E_j \cap E_k = \emptyset \text{ if } j \neq k, \ p_k \geq 0, \ \sum_{k=1}^{n} p_k = 1).$$

(Again, we may have also further sets of probabilities but those should be positive, at least when the corresponding $p$ is.) By *symmetry* we mean here that the value of $H_n$ does not change if two *columns*

$$E_j \quad \text{and} \quad E_k$$
$$p_j \quad\quad\quad p_k$$

are exchanged. The analogue of the recursity (7) is here

$$H_{n+1} \begin{pmatrix} E_1 \cap F_1, E_1 \cap F_2, E_2, \dots, E_n \\ p_1 q_1, \quad p_1 q_2, \quad p_2, \dots, p_n \end{pmatrix}$$

$$= H_n \begin{pmatrix} E_1, E_2, \dots, E_n \\ p_1, p_2, \dots, p_n \end{pmatrix} + p_1 H_2 \begin{pmatrix} F_1, F_2 \\ q_1, q_2 \end{pmatrix}$$

$$(E_j \cap E_k = \emptyset \text{ if } j \neq k, \ F_1 \cap F_2 = \emptyset,$$

$$p_k \geq 0, \ q_j \geq 0, \ \sum_{k=1}^{n} p_k = 1 = q_1 + q_2). \tag{10}$$

Again one may ask for the general form of symmetric measures satisfying (10) and, say, continuous in the probabilities. This has been proved [8,4], again with the aid of functional equations, to be (with the convention (3))

$$a \sum_{k=1}^{n} p_k \log p_k + \sum_{k=1}^{n} p_k g(E_k) - g(\bigcup_{k=1}^{n} E_k) \tag{11}$$

where $a$ is an arbitrary constant and $g$ an arbitrary real valued function on the ring of sets. Note that $\cup E_k = S$ is not supposed here ($S$ may not even be in the ring of sets considered). If we have Boolean rings (thus containing $S$) and

$$\bigcup_{k=1}^{n} E_k = S$$

then, with $h(E) = g(E) - g(S)$, (11) goes over into

$$a \sum_{k=1}^{n} p_k \log p_k + \sum_{k=1}^{n} p_k h(E_k). \tag{12}$$

(For the characterization of similar quantities depending upon *several* sets of probabilities, see [19,20].)

The quantities (11) and (12), consisting of Shannon's entropy plus additional terms, are called *inset entropies* of the Shannon type. (Inset may be understood in its dictionary meaning 'a map set into a map' or as 'in set', but really the name was chosen because the idea was born at a meeting at the Ecole Normale Supérieure de l'Enseignment Technique - ENSET - near Paris.)

There are applications of these measures in information theory and elsewhere. Traditionally

$$-\int_a^b f(x) \log f(x)\, dx \tag{13}$$

is considered to be the 'continuous' analogue of (1) for a random variable with probability density function $f$. Contrary to what one may think, the sums approximating (13)

$$-\sum_{k=1}^n f(\xi_k) \log f(\xi_k)(x_k - x_{k-1}) \tag{14}$$

$(a = x_0 \leq \xi_1 \leq x_1 \leq \xi_2 \leq \ldots \leq \xi_n \leq x_n = b)$ are *not* Shannon entropies of the form (1) but they *are* inset entropies of the form (12). We see this by realizing that here

$$p_k = f(\xi_k)(x_k - x_{k-1}), \quad E_k = (x_{k-1}, x_k],$$

$$\bigcup_{k=1}^n E_k = (a, b] \quad \text{and} \quad x_k - x_{k-1} = l(E_k)$$

($l$ standing for length), so that (14) goes over into

$$-\sum_{k=1}^n p_k \log p_k + \sum_{k=1}^n p_k \log l(E_k) \tag{15}$$

which is indeed of the form (12) [3].

However, contrary to (1), the quantity (13) and therefore also (15) may be negative. (That is one reason why (13) is not a very good analogue of the Shannon entropy (1) and also why in characterizations of the above inset entropies we have used continuity rather than boundedness – we certainly could not have used nonnegativity.) On the other hand, if $n=1$ and $E=(a,b]$, then $p_1=1$ and (15) reduces to $\log l(E)$. This corresponds to the case where we know that the value of the random variable falls into $(a,b]$ but don't know its probability distribution. Since

$$E(a,b] = \bigcup_{j=1}^{n} (x_{j-1},x_j] = \bigcup_{j=1}^{n} E_k,$$

the decrease of uncertainty between this ignorant state and the state described by (15), where we know the probability distribution, is [3]

$$\log l(\bigcup_{j=1}^{n} E_j) - \sum_{k=1}^{n} p_k \log l(E_k) + \sum_{k=1}^{n} p_k \log p_k, \qquad (16)$$

an inset entropy of the form (11) (this happens to be also nonnegative). If 0 probabilities are permitted and $p_i = 1$, $p_k = 0$ for $i \neq k$ (information that the value of the random variable falls into the $i$-th subinterval) then (16) reduces to the 'Wiener information measure'

$$-\log[l(E_i)/l(E)]. \qquad (17)$$

Also the

$$q_k = l(E_j)/l(\bigcup_{j=1}^{n} E_j)$$

can, of course, be considered (geometric) probabilities, therefore (17) may be written as $(-\log q_i)$ and (16) as $\Sigma p_k \log (p_k/q_k)$, a quantity which we have encountered before and called directed divergence. Conversely, since in inset entropies the dependence upon the events can appear as dependence upon *parameters* determined by the events, *all* directed divergences, information improvements, etc. may be considered inset entropies.

There have been recent efforts to take, in addition to the probabilities, also the 'usefulness' of events (again of weather, of market situations, etc.) into consideration. In view of the above it would seem worthwhile to look at these 'measures of useful information' as inset measures and thus develop an appropriate theory.

For a more playful application, to the theory of gambling, Meginnis [23] considers the *second* term in (12) the expected gain ($E_k$ being the $k$-th outcome of the game which has a gain in the amount $h(E_k)$ attached to it). So this time it is the *first* term which has to be explained. Since the expected gain alone would not motivate gambling (it is almost always nonpositive), he interprets the first term as quantifying the *joy in gambling*, (regardless whether it is a service or disservice to mankind). He too derives (12), with the aid of functional equations, from requirements fairly natural for this application.

Finally, the forecasting theory application, mentioned above, can also be naturally generalized to such inset situations. The payoff $f(E_k, q_k)$ may very well depend upon the $k$-th event itself, not only on its probability. Then the forecaster's expected gain is

$$\sum_{k=1}^{n} p_k f(E_k, q_k)$$

and we 'keep the forecaster honest' by choosing $f$ so that

$$\sum_{k=1}^{n} p_k f(E_k, q_k) \leq \sum_{k=1}^{n} p_k f(E_k, p_k). \qquad (18)$$

It has been proved [2] that the only solution of this functional inequality happens (for a fixed $n > 2$) when

$$f(E, q) = a \log q + h(E)$$

($a \geq 0$ an arbitrary constant, $h$ an arbitrary real valued function on the ring of events). So the right hand side of (18) goes over into

$$a \sum_{k=1}^{n} p_k \log p_k + \sum_{k=1}^{n} p_k h(E_k),$$

again an inset entropy of the form (12).

The characterization theory of inset measures of information is somewhat behind the probabilistic theory. Among others, as we have mentioned before, 0–probabilities have recently been eliminated from much of the characterization theory of purely probabilistic information measures. But in the similar theory of inset measures, impossible events (empty sets) are vigorously used. While the probabilities are separate variables, it seems to be common sense to require that impossible events have 0 probabilities. Up to now impossible events and 0 probabilities have been completely eliminated from (10) and its (symmetric) solutions (by Maksa and Ebanks in [34]) and in the solution of (18) in [2], mentioned above, keeping the expert honest with inset reward and without impossible events and zero probabilities (but still with some money). Some simple results in [5] and their extensions have been used here and may serve as tools for other inset characterization theorems without empty sets. This is still a wide *open domain* for further research.

In our opinion, precisely those serious applications of information theory to fields other than the classical communication theory could make good use of this new, mixed theory of information. Indeed, in the classical communication theory the *contents* of messages are usually ignored and only their frequencies, probabilities, etc. considered. This makes the probabilistic information theory the right tool there. But it seems to us that in the other applications mentioned above – and in many more – the contents are essential, and so it would be worth exploring whether the new theory could be more appropriately and efficiently applied there.

I did not aim at completeness in any issue (not even in the references), but I hope that I have succeeded to give at least an idea of some unconventional applications of information theory and also some food for thought concerning further work which seems worth doing.

Centre for Information Theory,
University of Waterloo,
Waterloo, Ont., Canada
N2L 3G1

## References

[1]   Aczél, J.: 1974, "Keeping the expert honest" revisited - or: A method to prove the differentiability of the solutions of functional inequalities. *Selecta Statist. Canad.* 2, 1-14.

[2]   Aczél, J.: 1980, A mixed theory of information. V. How to keep the (inset) expert honest. *J. Math. Anal. Appl.* 75, 447-453.

[3]   Aczél, J.: 1978-80, A mixed theory of information. VI. An example at last: A proper discrete analogue of the continuous Shannon measure of information (and its characterization). *Univ. Beograd. Publ. Elektrotehn. Fak. Ser. Mat. Fiz.* No. 602-633, 65-72.

[4]   Aczél, J.: 1980, A mixed theory of information. VII. Inset information functions of all degrees. *C.R. Math. Rep. Acad. Sci. Canada* 2, 125-129.

[5]   Aczél, J.: 1980, Functions partially constant on rings of sets. *C.R. Math. Rep. Acad. Sci. Canada* 2, 159-164.

[6]   Aczél, J.: 1980, Information functions on open domains. III. *C.R. Math. Rep. Acad. Sci. Canada* 2, 281-285.

[7]   Aczél, J. and Z. Daróczy: 1975, *On measures of information and their characterizations.* Academic Press, New York-San Francisco-London.

[8]   Aczél, J. and Z. Daróczy: 1978, A mixed theory of information. I. Symmetric, recursive and measurable entropies of randomized systems of events. *RAIRO Inform. Théor.* 12, 149-155.

[9]   Aczél, J. and C.T. Ng: Determination of all semisymmetric recursive information measures of multiplicative type on $n$ positive discrete probability distributions. *Linear Alg. and Appl. 52-53,* 1-30.

[10]  Attneave, F.: 1959, *Applications of information to psychology: A summary of basic concepts, methods, and results.* Henry Holt, New York.

[11]  Bruckmann, G.: 1969, Einige Bemerkungen zur statistischen Messung der Konzentration. *Metrika* 14, 183-213.

[12]  Cheng, M.C.: 1978, A justification for information functions as measure of detective performance. *Kybernetes* 7, 153-155.

[13]  Cowell, F.A.: 1980, Generalized entropy and the measurement of distributed change. *European Econom. Rev.* 13, 147-159.

[14]  Cowell, F.A. and K. Kuga: 1981, Inequality measurement - An axiomatic approach. *European Econom. Rev.* 15, 287-305.

[15]  Cowell, F.A. and K. Kuga: 1981, Additivity and the entropy concept: an axiomatic approach to inequality measurement. *J. Econom. Th.* 25, 131-143.

[16]  Gehrig, W.: 1983, On a characterization of the Shannon concentration measure. To appear in *Utilitas Math.*

[17]  Georgescu-Roegen, N.: 1971, *The entropy law and the economic process.* Harvard University Press, Cambridge, Mass.

[18]  Hardy, G.H.: 1967, *A mathematician's apology.* Cambridge University Press, Cambridge. Cambridge.

[19]  Kannappan, Pl.: 1980, A mixed theory of information. IV. Inset inaccuracy and directed divergence. *Metrika* 27, 91-98.

[20]  Kannappan, Pl. and W. Sander: 1982, A mixed theory of information. VIII. Inset measures depending upon several distributions. *Aequationes Math. 25,* 177-193.

[21]  Kolmogorov, A.N.: 1933, *Grundbegriffe der Wahrscheinlichkeitsrechnung.* Springer, Berlin, 1933.

[22]  Maksa, Gy. and C.T. Ng: 1983, The fundamental equation of information on open domain. To appear in *Publ. Math. Debrecen.*

[23]  Meginnis, J.R.: 1976, A new class of symmetric utility rules for gambles, subjective marginal probability functions and a generalized Bayes rule. *Bus. Econom. Statist. Sec. Proc. Amer. Statist. Assoc.* 471-476.

[24]  Ng, C.T.: 1980, Information functions on open domains. *C.R. Math. Rep. Acad. Sci. Canada* 2, 119-123.

[25]  Ng, C.T.: 1980, Information functions on open domains. II. *C.R. Math. Rep. Acad. Sci. Canada* 2, 155-158.

[26]  Pielou, E.: 1967, The use of information theory in the study of the diversity of biological populations. In: *Proceedings of the Fifth Berkeley Symposium on Mathematics Statistics and Probability,* Vol. IV, University of California Press, Berkeley, Calif., pp. 163-177.

[27]  Pielou, E.: 1977, *Mathematical ecology.* Wiley, New York-London-Sydney.

[28]  Piesch, W.: 1975, *Statistische Konzentrationsmasse.* Mohr, Tübingen.

[29]  Shannon, C.E.: 1956, The bandwagon. *IRE Trans. Inform. Theory* IT-2, 3.

[30]  Theil, H.: 1967, *Economics and information theory.* North Holland, Amsterdam - Rand McNally, Chicago.

[31]  Theil, H.: 1972, *Decomposition analysis (— with applications in the social and administrative sciences).* North Holland, Amsterdam-London.

[32]  Theil, H.: 1979, The measurement of inequality by components of income. *Econom. Lett.* 2, 197-199.

[33]  Willmer, M.A.P.: 1966, On the measurement of information in the field of criminal detection. *Oper. Res. Quart.* 17, 335-345.

[34]  The Twentieth International Symposium on Functional Equations, August 1-7, 1982, Oberwolfach, Germany (compiled by B. Ebanks). *Aequationes Math. 24,* 261-297, in part. pp. 269, 277.

W. Gehrig

## On a characterization of the Shannon concentration measure

## 1. Introduction

To the large class of functions that have been introduced in economic literature as measures of inequality (see [4,11,17]) there belongs one function which has its origin not in economics but in information theory, namely

$$S^n(\mathbf{x}) = \gamma \sum_{i=1}^{n} x_i \log x_i \quad (\gamma > 0 \ \text{const.}), \qquad (1.1)$$

where $0 \cdot \log 0 := 0$ and

$$\mathbf{x} \in \Gamma^n := \{\mathbf{y} \,|\, y_i \geq 0, \ y_1 + \ldots + y_n = 1\}.$$

In information theory $(-1/\gamma)S^n$ is the well known Shannon entropy (see [14,15]), the measure of information expected from finite complete probability distributions where the $x_1, \ldots, x_n$ are the probabilities of an experiment with outcomes $A_1, \ldots, A_n$. If (1.1) is used as a measure for the industrial concentration, $x_i$ is interpreted as the market share of the firm $i$. Many characterizations of the Shannon entropy based on different axioms, are known in information theory (see [2] for a comprehensive treatment of these papers), but, to the best of our knowledge, none is published that uses economically meaningful assumptions.

In Section 4 we give a characterization of (1.1) that is based upon the most obvious property which an inequality measure should possess, namely the so called strict principle of progressive transfers (see (T) in definition (2.1)) and a property called "additive decomposability" which formulates what happens if certain firms unite their market shares.

J. Aczél (ed.), Functional Equations: History, Applications and Theory, 191-205.
© 1984 by D. Reidel Publishing Company.

## 2. Definition of concentration measures

A definition of the notion "concentration measure" should be given in terms of "essential" and "natural" properties, which means, roughly speaking, properties that are economically significant and independent in the sense that none of these properties is a consequence of the others. We feel that the following definition satisfies these requirements.

DEFINITION (2.1). A concentration measure is a sequence of functions

$$\{K^n | K^n : \Gamma^n \to \mathbf{R}\},$$

such that

$$K^n(x_1,\dots,x_n) > K^n(x_1,\dots,x_i+h,\dots,x_j-h,\dots,x_n) \tag{T}$$

for all $i,j \in \{1,\dots,n\} : x_j > x_i$, for all $h \in (0, \frac{1}{2}(x_j-x_i)]$ and for all $\mathbf{x} \in \Gamma^n$;

$$K^n(P\mathbf{x}) = K^n(\mathbf{x}) \tag{S}$$

for all permutation matrices[1] $P \in \mathbf{P}$ and all $\mathbf{x} \in \Gamma^n$;

$$K^{n+1}(\mathbf{x};0) = K^n(\mathbf{x}) \tag{E}$$

for all $\mathbf{x} \in \Gamma^n$.

These properties are intepreted as follows:

First, the strict principle of progressive transfers (T) states that whenever a firm $(j)$ with a higher turnover looses a certain amount of turnover $(=h)$ to a firm $(i)$ with a smaller turnover, concentration should diminish; this property is also called the Pigou (see [12]) - Dalton (see [5]) - condition, since these two authors had been the first to introduce (T) into economic literature. Secondly, the symmetry (S) reflects the principle of "equal treatment of equals" or of "horizontal equality".

The extensibility (E) states that an additional firm with zero turnover does not change concentration.

These are the economic reasons for (T), (S) and (E). On the other hand, from the mathematical point of view, the properties (T), (S) and (E) are also well chosen since they are consistent and independent, as is shown in the following theorem.

THEOREM (2.2). The conditions (T), (S) and (E) are consistent. They are also independent in the following sense: For any two of them there exists a function $K^n: \Gamma^n \to \mathbf{R}$ that satisfies these two conditions but not the remaining third.

Proof. *Consistency:* The function

$$K^n(\mathbf{x}) = \sum_{i=1}^{n} x_i^2 \tag{2.3}$$

satisfies (T), (S) and (E).

*Independence:* The function

$$K^n(\mathbf{x}) = \sum_{i=1}^{n} x_i^2 - 1/n \tag{2.4}$$

satisfies (T) and (S), but not (E). The function

$$K^n(\mathbf{x}) = \begin{cases} 2 & \text{if } x=(1,0,...,0) \\ \sum_{i=1}^{n} x_i^2 & \text{otherwise} \end{cases} \tag{2.5}$$

satisfies (T) and (E), but not (S). The function

$$K^n(\mathbf{x}) = \sum_{i=1}^{n} x_i \tag{2.6}$$

satisfies (S) and (E), but not (T). Q.E.D.

## 3. Properties of concentration measures as defined in 2.1

The first result of this section shows for symmetric functions a fundamental relationship between the strict principle of progressive transfers (T) and the strict $S$-convexity[2,3].

THEOREM (3.1). The function $K^n : \Gamma^n \to \mathbb{R}$ is strictly $S$-convex if, and only if, it is symmetric (S) and satisfies the Pigou-Dalton-condition (T).

Proof. Every strictly $S$-convex function is symmetric (see [3, Theorem 3, p. 220]) and satisfies (T) (see [6]). This proves the "only if" part.

In order to prove the "if" part, we assume that $K^n$ is *not* strictly $S$-convex and show that this assumption yields a contradiction to (T). If $K^n$ is not strictly $S$-convex, then there exists an $\mathbf{x}^0 \in \Gamma^n$ and a $B^0 \in \mathbf{B}$ satisfying

$$K^n(B^0\mathbf{x}^0) \geq K^n(\mathbf{x}^0), \tag{3.2}$$

where the vector $\mathbf{y}^0 := B\mathbf{x}^0$ is not a permutation of $\mathbf{x}^0$. If we take the increasing rearrangement of both $\mathbf{x}^0$ and $\mathbf{y}^0$, the resulting vectors $\mathbf{x}^1$ and $\mathbf{y}^1$ satisfy

$$\mathbf{y}^1 = B\mathbf{x}^1 \quad (B\mathbf{x}^1 \text{ not a permutation of } \mathbf{x}^1) \tag{3.3}$$

and thus, by (3.3) and (S),

$$K^n(\mathbf{y}^1) \geq K^n(\mathbf{x}^0) \tag{3.4}$$

and, by (S),

$$K^n(\mathbf{x}^1) = K^n(\mathbf{x}^0). \tag{3.5}$$

Condition (3.3) is equivalent to the existence of a nonempty finite sequence of progressive transfers[4] that leads from $\mathbf{x}^1$ to $\mathbf{y}^1$ (see [6]):

$$\mathbf{x}^1 \overset{PT}{\rightarrow} \mathbf{x}^2 \overset{PT}{\rightarrow} \dots\dots \overset{PT}{\rightarrow} \mathbf{x}^k := \mathbf{y}^1. \tag{3.6}$$

Now we have derived a contradiction to (T), according to which

$$K^n(\mathbf{x}^1) > K^n(\mathbf{y}^1).$$

Q.E.D.

PROPOSITION (3.8). If a function $K^n : \Gamma^n \rightarrow \mathbb{R}$ is strictly quasiconvex[5] and symmetric, then it is strictly $S$-convex.

Proof. The definition of the strict quasiconvexity implies

$$K^n(\sum_{i=1}^{n} \lambda_i \mathbf{x}_i) < \max_i K^n(\mathbf{x}_i)$$

$$\mathbf{x}_1, \dots, \mathbf{x}_n \in \Gamma^n \quad (0 < \lambda_1 < 1; \ \lambda_1 + \dots + \lambda_n = 1) \tag{3.9}$$

for any arbitrary finite convex combination of the $\mathbf{x}_i$. Now, let $Q$ be an arbitrary bistochastic matrix that does not permute $\mathbf{x}$. As is shown in [3,p. 182], every such $Q$ can be written as

$$Q = q_1 P^1 + \dots + q_m P^m, \tag{3.10}$$

where the $P^j$ are permutation matrices; $m \geq 2$; $0 < q_i < 1$ and $\sum_{k=1}^{m} q_k = 1$.

From (3.9) and (3.10) we obtain for all such $Q$

$$\begin{aligned} K^n(Q\mathbf{x}) &= K^n((q_1 P^1 + \dots + q_m P^m)\mathbf{x}) \\ &= K^n(q_1(P^1\mathbf{x}) + \dots + q_m(P^m\mathbf{x})) \\ &< \max_i K^n(P^i\mathbf{x}) = K^n(\mathbf{x}), \end{aligned} \tag{3.11}$$

where the inequality sign is a consequence of the strict quasiconvexity and the last equality sign is a consequence of the symmetry. Q.E.D.

COROLLARY (3.12). If a function $K^n:\Gamma^n \to \mathbf{R}$ is strictly convex and symmetric, then it is strictly $S$-convex.


Proof. A strictly convex function is strictly quasiconvex. Q.E.D.


The following two results will be helpful in section 4.


THEOREM (3.13). If $K^n:\Gamma^n \to \mathbf{R}$ satisfies (T), then it also satisfies the "mean-value-test"

$$K^n(\frac{1}{n},...,\frac{1}{n}) \leq K^n(\mathbf{x}) \leq \max_{P \in P} K^n(P(1,0,..,0)) \qquad (3.14)$$

for all $\mathbf{x} \in \Gamma^n$ with strict inequality sign if $\mathbf{x} \neq (\frac{1}{n},...,\frac{1}{n})$ and if $\mathbf{x}$ is not a permutation of $(1,0,...,0)$.


Proof. Let $\mathbf{x} \in \Gamma^n$ be arbitrary. Then[4]

$$\mathbf{x} \xrightarrow{PT} ...... \xrightarrow{PT} (\frac{1}{n},....,\frac{1}{n}) \qquad (3.15)$$

and

$$\mathbf{x} \xrightarrow{RT} ...... \xrightarrow{RT} P(1,0,..,0) \text{ for one } P \in \mathbf{P}, \qquad (3.16)$$

where the respective sequence is empty, if either $\mathbf{x} = (\frac{1}{n},...,\frac{1}{n})$ or $\mathbf{x}$ is a permutation of $(1,0,.,0)$.


The inequalities (3.14) follow now from (T) by (3.15) and (3.16). Q.E.D.

PROPOSITION (3.17). If $K^n: \Gamma^n \to \mathbf{R}$ satisfies (T) and (E), then the function $\psi: \mathbf{N} - \{1\} \to \mathbf{R}$, defined by

$$\psi(n) := -K^n(\frac{1}{n}, \ldots, \frac{1}{n}) \quad (n=2,3,\ldots), \tag{3.18}$$

is strictly increasing.

Proof. The extensibility (E) yields

$$K^n(\frac{1}{n}, \ldots, \frac{1}{n}) = K^{n+1}(\frac{1}{n}, \ldots, \frac{1}{n}; 0), \tag{3.19}$$

whereas (T) implies, by (3.13),

$$K^{n+1}(\frac{1}{n}, \ldots, \frac{1}{n}; 0) > K^{n+1}(\frac{1}{n+1}, \ldots, \frac{1}{n+1}). \tag{3.20}$$

Q.E.D.

## 4. A characterization of the Shannon concentration measure by additive decomposability

Actually, the conditions (T), (S) and (E) do not uniquely determine a reasonable class of concentration measures. Therefore we have to add further properties in order to characterize special types of concentration measures. Before we introduce an additional condition, consider the following situation: Suppose that we aggregate the $n$ firms in such a way that $l$ sets $S_1, \ldots, S_l$ are created. Each firm belongs to exactly one such set. We write $n_g$ for the number of firms in $S_g$, so that $n_1 + \ldots + n_l = n$. Then the following equation, which we call "additive decomposability" (see [16,p.93]), should hold

$$K^n(\mathbf{x}) = K^l(X_1, \ldots, X_l) + \sum_{j=1}^{l} X_j K^{|n_j|}(\frac{\mathbf{x}^j}{X_j}) \tag{D}$$

for all $\mathbf{x} \in \Gamma^n$, $n \geq 3$, with the convention

$$0 \cdot K^{|n_j|}(\frac{0}{0}) := 0 \quad (j=1,2,\ldots,l).$$

In (D), $X_j := \sum_{i \in S_j} x_i$ and the vector $\mathbf{x}^j$ consists of the market shares that are held by firms belonging to $S_j$. Note that (D) implies $K^1(1)=0$.

The following lemma shows that, in what follows, we can omit (E).

LEMMA (4.1). If $K^n:\Gamma^n \to \mathbf{R}$ satisfies (S) and (D), then $K^n$ also satisfies (E).

Proof. From (D) we obtain, with $l=n-1$, $S_1=\{1,2\}$ and $S_i=\{i+1\}$ $(i=2,..,n)$,

$$K^n(\mathbf{x}) = K^{n-1}(x_1+x_2,x_3,..,x_n)$$

$$+ (x_1+x_2)K^2\left(\frac{x_1}{x_1+x_2}, \frac{x_2}{x_1+x_2}\right), \qquad (R)$$

a property that is called "recursivity" in information theory (see [2,p.51]). Proposition (2.35) in [2,p.59] establishes now the proof. Q.E.D.

In the proof of Theorem (4.20) the function $f:[0,1] \to \mathbf{R}$ defined by

$$f(x) := K^2(x,1-x) \qquad (4.2)$$

will play a certain role. Therefore we first derive some properties of $f$ that follow out of (T), (S) and (R).

LEMMA (4.3). Let $K^n:\Gamma^n \to \mathbf{R}$ satisfy (T), (S) and (R). Then the function $f$, as defined by (4.2), has the following properties:

(i)  $f$ satisfies the functional equation

$$f(x)+(1-x)f(\frac{y}{1-x}) = f(y)+(1-y)f(\frac{x}{1-y}) \qquad (4.4)$$

for all $x,y \in [0,1)$, $x+y \in [0,1]$;

(ii)  $f$ is bounded on $[0,1]$ with

$$f(1/2) \leq f(x) \leq \max\{f(0),f(1)\}; \qquad (4.5)$$

(iii)

$$f(x) = f(1-x); \qquad (4.6)$$

(iv)  $f$ is nonpositive throughout and satisfies

$$f(0) = f(1) = 0; \qquad (4.7)$$

(v)  $f$ is strictly decreasing on $[0,1/2]$ and strictly increasing on $[1/2,1]$;

(vi)

$$f \in C^{\infty}([0,1]). \qquad (4.8)$$


Proof.

(i)  See [2,p.77].

(ii)  By (3.13)

$$K^2(1/2,1/2) \leq K^2(x,1-x)$$
$$\leq \max\{K^2(0,1),K^2(0,1),\}, \qquad (4.9)$$

which means, by (4.2),

$$f(1/2)\leq f(x)\leq \max\{f(0),f(1)\}. \qquad (4.10)$$

(iii)  Direct consequence of (S) and (4.2).

(iv)  From (R) we obtain in case $n=3$, by setting $x_1=x_2=1/2$ and $x_3=0$,

$$K^3(\frac{1}{2},\frac{1}{2},0) = K^2(1,0)+K^2(\frac{1}{2},\frac{1}{2}), \qquad (4.11)$$

and, by setting $x_1 = x_3 = 1/2$ and $x_2 = 0$,

$$K^3(\frac{1}{2},0,\frac{1}{2}) = K^2(\frac{1}{2},\frac{1}{2})+\frac{1}{2}K^2(1,0). \tag{4.12}$$

By (S), the left-hand sides of (4.11) and (4.12) are equal, and so are the right-hand sides, from where we obtain

$$K^2(1,0) = \frac{1}{2}\, K^2(1,0), \tag{4.13}$$

which is possible if, and only if, $K^2(1,0)=0$. Now (iv) follows from (4.5).

(v)   Direct consequence of (T) by (4.2).

(vi)  In proving (4.8), we use a method that is sometimes applied in functional equation theory in order to derive from weaker regularity conditions (such as measurability or integrability) stronger ones (such as continuity or differentiability) (see [1,p.190] or [9,p.417]).
From (v) we know that $f$ is (Riemann)-integrable over every closed interval $[a,b]\subset[0,1)$. Now, let $y\in[0,1)$ be arbitrary, but fixed. We integrate (4.4) over $[0,b]$ $0<b<1;b+y<1)$ in the sense of Riemann with respect to $x$ and obtain

$$b\cdot f(y) = \int\limits_0^b f(x)\,dx+\int\limits_0^b (1-x)f(\frac{y}{1-x})\,dx$$

$$- (1-y)\cdot\int\limits_0^b f(\frac{x}{1-y})\,dx \tag{4.14}$$

From the well-known formula (see [8,p.30])

$$\int\limits_a^b f(x)\,dx = \int\limits_{\phi^{-1}(a)}^{\phi^{-1}(b)} f(\phi(t))\phi'\,(t)\,dt \tag{4.15}$$

we obtain

$$b\cdot f(y) = \int\limits_0^b f(x)\,dx + y^2 \int\limits_y^{y/(1-b)} s^{-3}f(s)\,ds$$

$$- (1-y)^2 \int\limits_0^{b/(1-y)} f(t)\,dt. \tag{4.16}$$

The right-hand side of (4.16) is continuous in $y$ (on $[0,1)$) and so is $f$ on the left-hand side. Now, $f \in C^\infty([0,1))$ follows by iteration and (4.8) by (S). Q.E.D.

Before we prove our main theorem, we give in the class of recursive (R) and symmetric (S) functions a characterization of all functions satisfying (T).

LEMMA (4.17). Let $K^n : \Gamma^n \to \mathbb{R}$ satisfy (R) and (S). Then $K^n$ satisfies (T) if, and only if, $p \to K^2(p, 1-p)$ is strictly decreasing on $[0, 1/2]$.

Proof. The "only if" part follows from (4.3) without assuming (R) or (S).

As to the "if" part, we consider

$$K^n(\mathbf{x}) = K^{n-1}(x_1 + x_2, x_3, \ldots, x_n)$$

$$+ (x_1 + x_2) K^2\left( \frac{x_1}{x_1 + x_2}, \frac{x_2}{x_1 + x_2} \right). \tag{R}$$

Now assume, without loss of generality, that $0 \le x_1 < x_2$. We replace $x_1$ by $x_1 + h := x_1'$ and $x_2$ by $x_2 - h := x_2'$ with $h \in (0, \frac{1}{2}(x_2 - x_1)]$. Then

$$x_1 + x_2 = x_2', \quad \frac{1}{2} \ge \frac{x_1'}{x_1' + x_2'} > \frac{x_1}{x_1 + x_2}$$

$$\text{and} \quad \frac{x_2'}{x_1' + x_2'} < \frac{x_2}{x_1 + x_2}. \tag{4.18}$$

Since

$$K^2\left( \frac{x_1}{x_1 + x_2}, \frac{x_2}{x_1 + x_2} \right) > K^2\left( \frac{x_1'}{x_1' + x_2'}, \frac{x_2'}{x_1' + x_2'} \right), \tag{4.19}$$

we obtain

$$K^n(x_1, x_2, x_3, .., x_n) > K^n(x_1+h, x_2-h, x_3, ., x_n)$$

for all $h \in (0, \frac{1}{2}(x_2-x_1)]$ $(x_2 > x_1)$, i.e., $K^n$ satisfies (T) for the pair $(i,j)=(1,2)$. The symmetry (S) then concludes the proof. Q.E.D.

We are now ready to prove the following.

THEOREM (4.20). A function $K^n$: $\Gamma^n \to \mathbb{R}$ satisfies (T), (S) and (D) if, and only if,

$$K^n(\mathbf{x}) = \gamma \sum_{i=1}^n x_i \log x_i, \qquad (4.21)$$

where $\gamma > 0$ is a real constant and $0 \cdot \log 0 := 0$.

Proof. That (4.21) satisfies (S) and (D) is obvious. But (4.21) also satisfies (T) since $K^n$ is strictly $S$-convex (see [10,p.71]). This proves the "if" part.

As to the "only if" part, we consider the function $\phi: \mathbb{N} \to \mathbb{R}$ defined by

$$\phi(n) := \begin{cases} 0 & \text{if } n=1 \\ -K^n(\frac{1}{n}, .., \frac{1}{n}) & \text{otherwise.} \end{cases} \qquad (4.22)$$

The property (D) implies (R) and therefore $K^2(1,0)=0$ (see (4.3 iv) and (4.2)). But (D) and (S) imply (E) (see (4.1)), so that

$$K^n(1,0,...,0) = 0. \qquad (4.23)$$

From (4.23), (S) and (T) we obtain (see Lemma (3.13))

$$K^n(\mathbf{x}) \begin{cases} =0 & \text{if } \mathbf{x}=P(1,0,..,0), \ P \in \mathbb{P} \\ <0 & \text{otherwise.} \end{cases} \qquad (4.24)$$

By (4.23) and (4.24) we have

$$\phi(n) \begin{cases} =0 & \text{for } n=1 \\ >0 & \text{for } n\geq 2. \end{cases} \qquad (4.25)$$

But $\phi$ satisfies also the inequality (see (3.17))

$$\phi(n+1) > \phi(n) \quad \text{for all } n\in \mathbb{N}. \qquad (4.26)$$

On the other hand, it is well known that (R) and (S) yield

$$\phi(nm) =\phi(n) + \phi(m) \quad (n,m\in N) \qquad (4.27)$$

(see [2,p.65]). From (4.26) and (4.27) we obtain (see [7])

$$\phi(n) = \gamma\log n \quad (\gamma>0; n\in \mathbb{N}). \qquad (4.28)$$

Using similar arguments as in [13], it follows for all positive rationals $r \in (0,1)$ that

$$f(r) = \gamma(r \log r + (1-r) \log(1-r)) \qquad (4.29)$$

from where we obtain, by the regularity conditions on $f$ (see (4.3)),

$$K^2(x,1-x) = f(x) = \gamma(x \log x + (1-x) \log(1-x))$$

$$\text{for all } x\in[0,1]. \qquad (4.30)$$

This is (1.1) for $n=2$. Finally, the recursivity (R) extends (4.30) to arbitrary $n>2$. Q.E.D.

Allianz-Lebensvericherungs -AG
Reinsburgstr. 19
D-7000 Stuttgart
W. Germany

## Notes

1. A square matrix $P=(p_{ij})$ is said to be a *permutation matrix* if each row and column has one single unit and all other entries are zero. We denote the set of all permutation matrices by **P**.

2. An $n\times n$-matrix $B=(b_{ij})$ is called *doubly stochastic* or *bistochastic* if

$$b_{ij} \geq 0 \quad \text{and} \quad \sum_j b_{ij} = \sum_i b_{ij} = 1 \quad (i,j=1,..,n).$$

We denote the set of all bistochastic matrices by **B**. According to a theorem of Birkhoff and von Neumann (see [3,p.182]) **B** is the convex hull of **P**.

3.  A function $F{:}\Gamma^n{\to}\mathbf{R}$ is said to be *strictly S-convex* if

$$F(B\mathbf{x}) \leq F(\mathbf{x})$$

for all $B \in \mathbf{B}$ and all $\mathbf{x} \in \Gamma^n$ with

$$F(B\mathbf{x}) < F(\mathbf{x})$$

whenever $B\mathbf{x}$ is *not* a permutation of **x**. Note that the last condition is *not* equivalent to the condition that $B \notin \mathbf{P}$.

4.  The notation $\mathbf{x}^i \overset{PT}{\to} \mathbf{x}^{i+1}$ means that $\mathbf{x}^{i+1}$ is obtained from $\mathbf{x}^i$ by one single *progressive* transfer.

The notation $\mathbf{x}^l \overset{RT}{\to} \mathbf{x}^{l+1}$ means that $\mathbf{x}^{l+1}$ is obtained from $\mathbf{x}^l$ by one single *regressive* transfer (i.e. a transfer from "poor" to "rich").

5.  A function $F{:}\ \Gamma^n{\to}\mathbf{R}$ is called *strictly quasiconvex* if

$$F(\gamma\mathbf{x}+(1{-}\gamma)\mathbf{y}) < \max\{F(\mathbf{x}),F(\mathbf{y})\}$$

for all $\gamma{\in}(0,1)$ and all $\mathbf{x},\mathbf{y}{\in}\Gamma^n$ with $\mathbf{x}{\neq}\mathbf{y}$.

## References

[1]  Aczél, J.: 1966, *Lectures on functional equations and their applications*, Mathematics in Science and Engineering, Vol. 19, Academic Press, New York-London.

[2]  Aczél, J. and Z. Daróczy: 1975, *On measures of information and their characterizations*, Mathematics in Science and Engineering, Vol. 115, Academic Press, New York-San Francisco-London.

[3]  Berge, C.: 1963, *Topological spaces*, (English Translation), Oliver and Boyd, Edinburgh.

[4]  Bruckmann, G.: 1969, Einige Bemerkungen zur statistischen Messung der Konzentration. *Metrika* 14, 183-213.

[5]   Dalton, H.: 1920, The measurement of the inequality of incomes. *Economic J.* 30, 348-361.

[6]   Dasgupta, P., A. Sen and D. Starrett: 1973, Notes on the measurement of inequality. *J. of Economic Theory* 6, 180-187.

[7]   Erdös, P.: 1946, On the distribution function of additive functions. *Ann. of Math.* 47, 1-20.

[8]   Erwe, F.: 1973, *Differential- und Integralrechnung*, Band 2, B.I. Hochschultaschenbücher, Band 31, Bibliograpisches Institut, Mannheim.

[9]   Lee, P.M.: 1964, On the axioms of information theory. *Ann. Math. Statist.* 35, 415-418.

[10]  Marshall, A.W. and J. Olkin: 1979, *Inequalities: Theory of majorization and its applications.* Mathematics in Science and Engineering, Vol. 143, Academic Press, New York.

[11]  Piesch, W.: 1975, *Statistische Konzentrationsmasse.* J.C.B. Mohr, Tübingen.

[12]  Pigou, A.C.: 1912, *Wealth and welfare,* MacMillan, London.

[13]  Re͂nyi, A.: 1962, *Wahrscheinlichkeitsrechnung. Mit einem Anhang über Informationstheorie.* Deutscher Verlag der Wissenschaften, Berlin.

[14]  Shannon, C.E.: 1948, A mathematical theory of communication. *Bell System Tech. J.* 27, 379-423.

[15]  Shannon, C.E.: 1948, A mathematical theory of communication. *Bell System Tech. J.* 27, 623-656.

[16]  Theil, H.: 1967, *Economics and information theory.* North Holland Publishing Company, Amsterdam, 1967.

[17]  Wagenhals, G.: 1981, *Wohlfahrtstheoretische Implikationen von Disparitätsmass.* Mathematical Systems in Economics, Vol. 60, Verlagsgruppe Athenäum/Hain/Scriptor/Hanstein, Königstein (Taunus).

R. Thibault

# Closed invariant curves of a noncontinuously differentiable recurrence

The determination of invariant curves (when they exist) is a key problem in the study of the phase portrait of a second order autonomous recurrence (cf. [1])

$$x_{n+1} = g(x_n, y_n) \quad y_{n+1} = f(x_n, y_n). \tag{1}$$

In general, this determination proceeds by maximizing the regularity of a curve joining a phase plane point to its iterate (cf. sec. 2.1 of [1]); one is thus led to look for at least locally analytical solutions $\phi(x, y)$ of the functional equation

$$\phi(g(x, y), f(x, y)) = \phi(x, y) \tag{2}$$

or one of its equivalents. The interest in invariant curves of nonsmooth recurrences may thus appear paradoxical, especially if these curves possess angular points.

Such invariant curves appear nevertheless naturally in the study of the recurrence

$$T: x_{n+1} = y_n \quad y_{n+1} = -x_n + F(y_n) \tag{3}$$

when $F(y)$ is a continuous function, continuously differentiable for $y \neq 0$ and possessing a discontinuity of slope at $y=0$. The study of the particular case $F(y) = \lambda |y|$ has shown ([2]) that for certain values of $\lambda$ the origin of the phase plane is surrounded by continuous closed invariant curves, consisting of arcs of conic sections ("cyclic arcs") which are joined at angular points $P_i$ ($i=1,2,...,N$). These arcs are mutually interchanged by $T$ and its integer powers $T^m$. The angular points $P_i$ are the successive iterates of one of them, $P_1$, which is itself the iterate of a point $P_0$ situated on the "transition line" $y=0$. The slope of the ordinary tangent at $P_0$ is so mapped by the two different determinations of $T$, for $y>0$ and $y<0$, onto

J. Aczél (ed.), Functional Equations: History, Applications and Theory, 207-215.
© 1984 by D. Reidel Publishing Company.

two distinct half-slopes at $P_1$. The opposite behaviour occurs for $P_N$, the last angular point, which is situated on the transition line: the two distinct half-slopes at $P_N$ are mapped onto the same slope at $P_{N+1}$, restoring the regularity of the curve. Figure 1 shows the case $\lambda = 2^{1/4}$, corresponding to $N=8$. It has been shown in [2] why the two linear transformations $T_1$ and $T_2$, which are the two determinations of $T$, map the different arcs of ellipses onto themselves: in this example, $T_1$ and $T_2 T_1^6 T_2$ are equivalent in the sense that invariant curves associated with these two linear transformations are the same.

The first goal of this paper is to give a classification of cyclic arcs. It is natural, as for ordinary cycles, to introduce the following characteristic numbers.

1. The order $N$, which is now the number of the angular points of the invariant curve.

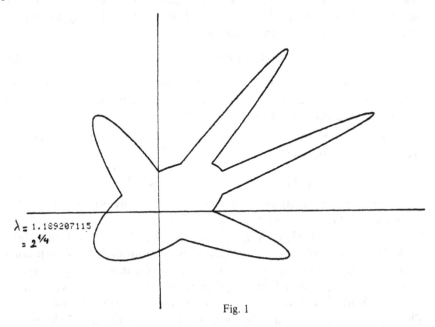

$\lambda = 1.189207115$
$= 2^{1/4}$

Fig. 1

2. The number of rotation $r$ involved in the mapping $P_0 \to P_N$. This number is necessarily of the form $(2k+1)/2$, if it is expressed in turns, because $P_0$ and $P_N$ belong to two different half-axes $\Delta_0$ $(y=0, x<0)$ or $\Delta_1$ $(y=0, x>0)$.

3. The last criterion is the species, defined to be equal to 1 if the invariant curve is composed of arcs of ellipses, or 2 if it is composed of arcs of hyperbolas. In the case mentioned above, $F(y)=\lambda|y|$, and the species is straightforwardly determined by the following remark: the species is 1 if $P_0$ belongs to $\Delta_0$ and $P_N$ belongs to $\Delta_1$, or 2 in the opposite case (Fig. 2).

The values of $N$ corresponding to a given value of $r$ are greater than $4r$ for the species 1 and greater than $4r-1$ for the species 2. But not all these values of $N$ are acceptable: when the corresponding value of $\lambda$ belongs to a divergence interval (see [3]), only isolated points are obtained. This is the case for $r=3/2$, $N=9,11,\ldots$ (species 1). The ratio $N/r$ is a monotonic function of $\lambda$ for each species (but not for their union). The numbers $N$ and $r$ being given, $\lambda$ is determined by algebraic equations involving Chebyshev polynomials.

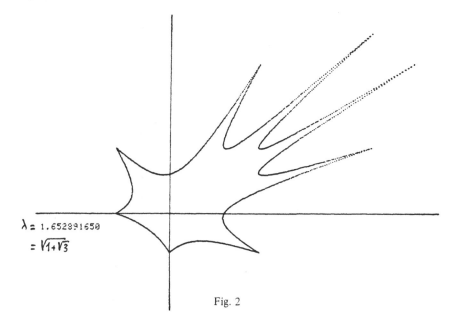

$\lambda = 1.652391650$

$= \sqrt{1+\sqrt{3}}$

Fig. 2

Sequences of such curves are interesting to study, for example, the sequence $N=4r+1$ (species 1) and $N=4r$ (species 2) for which the limiting value is 0. For $r=(4k-1)/2$ and $N=9k-2$ or $N=9k-1$, the sequence corresponds to values of $\lambda$ whose limit is 1 (species 1); the similar case for the second species is $r=(4k+1)/2$ and $N=9k+1$ or $N=9k+2$ (Fig. 3).

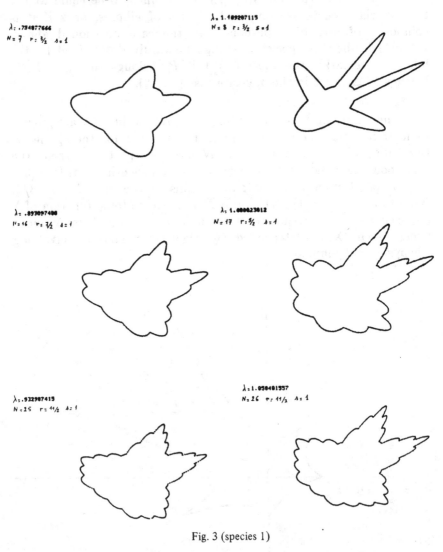

Fig. 3 (species 1)

$\lambda = .786151370$
$N = 10 \quad r = 5/2 \quad A = 2$

$\lambda = 1.123732021$
$N = 11 \quad r = 5/2 \quad s = 2$

$\lambda = .906726475$
$N = 15 \quad r = 9/2 \quad A = 2$

$\lambda = 1.673185927$
$N = 20 \quad r = 9/2 \quad A = 2$

$\lambda = .939333721$
$N = 28 \quad r = 13/2 \quad A = 2$

$\lambda = 1.051839256$
$N = 29 \quad r = 13/2 \quad A = 2$

Fig. 3 (species 2)

## Generalization

The function $F$ is now defined by $F(y)=\lambda|y|+\mu y$. Setting $A=\lambda+\mu$ and $B=\mu-\lambda$, the two determinations of $T$ are:

$$T_1:\quad x_{n+1}=y_n \quad y_{n+1}=-x_n+Ay_n \quad \text{for} \quad y_n>0$$

$$T_2:\quad x_{n+1}=y_n \quad y_{n+1}=-x_n+By_n \quad \text{for} \quad y_n<0.$$

It is noted that the symmetry $x\rightarrow-x$, $y\rightarrow-y$ permits us to exchange $A$ and $B$. The classification of cyclic arcs may be maintained, $P_0$ belonging now always to $\Delta_0$. The existence of arcs of ellipses and arcs of hyperbolas is still noticeable. Furthermore, there exist cyclic arcs whose number of rotation is $1/2$ (Fig. 4). Effective construction may be made as follows.

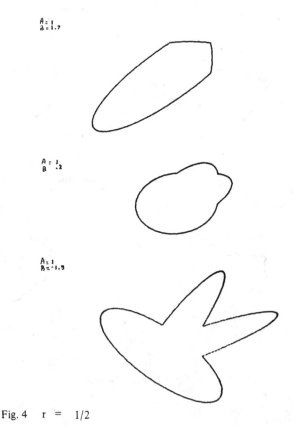

Fig. 4   r = 1/2

Starting with the point $(-1,0)$, one can exhibit, until a given order, all the possible cyclic arcs by writing the condition (a) $y_n = y_{n+1}$ or (b) $y_n = y_{n+2}$, taking into account the symmetry in the phase plane of $T$ with regard to the first bissectrix: (a) gives the cyclic arcs of odd order, (b) those of even order. Explicit relations between $A$ and $B$ may be deduced from the following table.

| $x_0$ | $y_0$ | $y_1$ | $y_2$ | $y_3$ | $y_4$ | $y_5$ |
|---|---|---|---|---|---|---|
| −1 | 0 | 1 | $A>0$ | $A^2-1>0$ | $A^3-2A>0$ | $A^4-3A^2+1$ |
| | | | | | $A^3-2A<0$ | $A^3B-2AB-A^2+1$ |
| | | | | | $A^2B-A-B$ | $A^3B-2A^2-AB+1$ |
| | | | | $<0$ | $>0$ | |
| | | | | | | $A^2B^2-AB-A^2-B^2+1$ |
| | | | | | $<0$ | $\cdots$ |
| | | | $<0$ | $AB-1>0$ | $A^2B-2A$ | |
| | | | | $<0$ | $AB^2-A-B$ | |

We classify now with respect to the order $N$ (until $N=6$)

| $N$ | $r$ | Values of $A$ and $B$ | Species or Particularity |
|---|---|---|---|
| 3 | 1/2 | $A=1$ | 1 |
| 4 | 1/2 | $A=\sqrt{2}$ | 1 or 2 |
| 4 | 3/2 | $AB=2$ | only exceptional cycles |
| 5 | 1/2 | $A=2\cos\pi/5$ | 1 or 2 |
| 5 | 3/2 | $B=1+1/A$ | 1 or 2 |
| 6 | 1/2 | $A=\sqrt{3}$ | 1 or 2 |
| 6 | 3/2 | $B=2A/(A^2-1)$ | 1 or 2 |
| 6 | 3/2 | $A=B/(B^2-2)$ | 1 |
| 6 | 5/2 | $AB=3$ | only exceptional cycles |

Transition between species 1 and 2 happens when the following condition is fulfilled. Let $P_i, P_j$ be the two angular points whose polar angles are nearest to $\pi$ and whose ordinates are of distinct signs. The condition is that the three points $P_0, P_i$ and $P_j$ fall on a line. In some cases, the transition involves the appearance of exceptional cycles (i.e. cycles whose points are indeterminate, because a power of $T$ becomes equal to the identity). For example, if $B=1+1/A$, the transition is obtained for $A=-1$, $B=0$, and in this case $T^7=I$. In other cases, the transition allows continuous passage from elliptic arcs to hyperbolic arcs by a polygon. An example is shown in Figure 5 for $N=7$, $r=5/2$, $B=(2A-1)/(A^2-A)$, the transition occurring for $A=(1+\sqrt{5})/2$, $B=-1$.

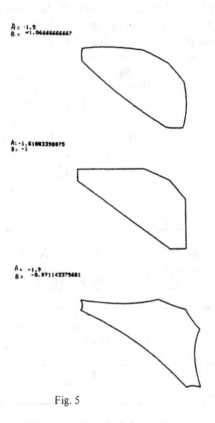

A = -1.5
B = -1.06666666667

A = -1.61803398875
B = -1

A = -1.5
B = -0.871143375681

Fig. 5

Obviously, the set of values of $(A,B)$ given in the above table, is to be restricted to the exterior of the domain of divergence (see [3]). For example, $B=1+1/A$ gives cyclic arcs of order 5 only if $A<1-\sqrt{2}$.

Mathématiques
Université Paul Sabatier
3 rue Edmond Rostand
F-31170 Tournefeuille
France

## References

[1]  Gumowski, I. and Ch. Mira: 1980, *Recurrences and discrete dynamic systems.* Lecture Notes in Math. Vol. 809, Springer Verlag, Berlin-New York.

[2]  Thibault, R.: 1981, Etude d'une récurrence positivement homogène. *C.R. Acad. Sci. Paris 292*, 225-230.

[3]  Thibault, R.: 1981, Influence of derivative discontinuities in second order recursive equations. *International Congress on Nonlinear Oscillations*, Kiev.

R.L. Clerc and C. Hartmann

**Invariant curves as solutions of functional equations**

## 1. Introduction.

Numerous evolution processes observed in nature are modellized by discrete dynamic systems, which can be interpreted as recurrences or point mappings. One of the simpler $\mathbf{R}^2 \to \mathbf{R}^2$ examples is (see a.o. [3], [5], [7], [8])

$$T: \quad (x,y) \to (X = g(x,y), \quad Y = f(x,y)) \tag{1}$$

where $f, g$ are smooth functions of their arguments. The singularities of (1) are necessarily of dimension zero (fixed points and cycles of a finite order $k \geq 1$) and one (invariant curves and analoguous invariant manifolds). Let $w(x,y) = \text{const.} = \alpha$ be the equation of an invariant curve ($\alpha$ fixed), then $w(x,y)$ is an automorphic function satisfying the functional equation (see for example [1], [3])

$$w(g(x,y), f(x,y)) = w(x,y) \tag{2}$$

Many point mappings of form (1) are known to possess an isolated closed invariant curve. Although most of them are noninvertible endomorphisms (examples: [4], [5], [7]), some happen to be at least local diffeomorphisms (example: [8]).

Isolated closed invariant curves can come into existence by several bifurcations, but the simplest one appears to be the destabilization of a fixed point or of a cycle focus analoguous to the Poincaré (-Hopf) bifurcation of the theory of autonomous second order differential equations.

Assume that $f, g$ in (1) depend also on a parameter $c$ and let $c_0$ be the corresponding bifurcation value. The bifurcated closed invariant curve $\mathbf{C}$ is known to be analytical only in a small neighbourhood of $c_0$, i.e. inside the parameter interval

217

*J. Aczél (ed.), Functional Equations: History, Applications and Theory, 217-226.*
© *1984 by D. Reidel Publishing Company.*

$]c_0, c_0+\epsilon[$, $0 < \epsilon \ll 1$. What happens outside of this interval is still a subject of research. The theorems on local existence of analytic invariant curves **B** crossing a nondegenerate saddle or node (eigenvalues $s \neq 0, \neq \pm 1$) [2] provide a promising method of investigation. This segment **B** can be extended by iteration by the use of (1) without losing its analyticity (at least for a finite number of iteration steps $m$). It may turn out that **B** approaches **C** asymptotically (by winding around it) as $m$ increases indefinitely $(m \rightarrow \infty)$. The properties of **C** can thus be studied by examining the behaviour of **B**.

## 2. Invariant Curve: Asymptotic Limit of an Unstable Submanifold

Consider a mapping $T$ of type (1) possessing at least two isolated fixed points: a saddle $S$ and a focus at the origin 0 which becomes unstable for a value $c_0$ of the parameter $c$ ($c$ is growing). If $S(s_0, y_0)$ denotes such a saddle (of eigenvalues $s \neq 0, \neq \pm 1$), it is known [2] that there exist two invariant arcs, one of consequents and one of antecedents with equations $y = \psi^{\pm}(x)$, defined for $|x - x_0|$ sufficiently small. Here $\psi^+$ and $\psi^-$ are two analytic functions which are solutions of the functional equations

$$f(x, \psi^+(x)) = \psi^+(g^+(x, \psi^+(x))), \qquad (3+)$$

or

$$f(x, \psi^-(x)) = \psi^-(g^-(x, \psi^-(x))), \qquad (3-)$$

respectively, where $g^-, f$ define locally the inverse mapping of $T$. If $T$ is a diffeomorphism, the continuation (without loss of analyticity) of the arcs $y = \psi^{\pm}(x)$ can be done without theoretical difficulties, at least as far as the definitions of $g^-, f$ are concerned (see [3]). Otherwise, when $T$ is a noninvertible endomorphism, which is now supposed, the continuation of $y = \psi^-(x)$ requires the knowledge of the inverse branches $T_{\epsilon_i}^{-1}$, $\epsilon_i = 1, ..., N$ of $T$ and of the critical curves $J_p$ of $T$

$$J_0 = \{M \in \mathbb{R}^2; \, J(M) = 0\}, \quad J_p = T^p J_0, \quad p \in \mathbb{N}$$

and of their antecedents (cf. [6]).

Since $T$ is now an endomorphism, the nondegenerate saddle $S$ belongs necessarily to the intersection of the domain of definition $\mathbf{D}_{\epsilon_{,}}$ of a mapping $T_{\epsilon_{,}}^{-1}$ (that defined by $g^{-}, f$ in (3–) or that which satisfies $T_{\epsilon_{,}}^{-1}(S) = S$) and of its image $T_{\epsilon_{,}}^{-1}(\mathbf{D}_{\epsilon_{,}})$

The continuation of the arc $y = \psi^{-}(x)$ permits us to define, if it exists, the bounded domain of no escape $D$ of $T$ (see [6]). Consider the case where $D$ exists and is simply connected for an interval of parameter values containing $c_0$. Let $\mathbf{B}_{\mathrm{I}}^{+}$ be the continuation of the arc of $y = \psi^{+}(x)$ interior to $D$, this arc being obtained as an asymptotic limit (as $m \to \infty$) of the iterate of order $m$ by $T$ of the entering arc of $y = \psi^{+}(x)$.

For $c < c_0$, $\mathbf{B}_{\mathrm{I}}^{+}$ approaches asymptotically (by winding around it) the focus 0, at least when 0 is the only stable singularity in $D$.

For $c > c_0$, if one observes that $\mathbf{B}_{\mathrm{I}}^{+}$ no longer has an asymptotic limit-point, it is possible to affirm that the destabilization of the focus 0 has produced, in the neighbourhood of 0, a stable submanifold of dimension 0 or 1, by a Poincaré-type bifurcation: an isolated closed invariant curve $\mathbf{C}$, or a pair $C_k$ of cycles of a finite order $k$, stable node-saddle, or any other strange attractor (this does not prevent the appearance elsewhere of other stable singularities in $D$). In the case of cycles $C_k$, it will be possible to construct by the preceding method the arcs $\mathbf{B}_k^{+}$ starting from saddles and, if the latter have as asymptotic limit points the corresponding stable nodes, then they will form a singular closed invariant curve.

Thus the existence of the stable submanifold $\mathbf{C}$ (or $C_k$) is assured as an asymptotic limit of the singular unstable submanifold $\mathbf{B}_{\mathrm{I}}^{+}$ (and of $\mathbf{B}_k^{+}$ in the case of $C_k$), and, at least for the case of $\mathbf{C}$, the continuation of $y = \psi^{+}(x)$, a solution of (3+), furnishes an asymptotic envelope of $\mathbf{C}$ and permits us to define an automorphic function which is a solution of (2).

For increasing values of the parameter, the curve $\mathbf{B}_{\mathrm{I}}^{+}$ may show self-intersection points and then cut $\mathbf{C}$ at nondegenerate (and nonstochastic) heteroclinic points. As far as the curve $\mathbf{C}$ is concerned, it may lose a certain number of good properties as soon as it has some common points with the critical curves $J_p$ of $T$.

## 3. Application to a Quadratic Endomorphism

Consider the particular case of an endomorphism $T$, of class at least $C^1$, such that

(i)   $J_0$ has just a single branch

(ii)  $J_1$ separates the plane $\mathbf{R}^2$ into two disjoint regions $\Sigma_2$ (or $\Sigma_0$) of two (or 0) antecedents: $\mathbf{R}^2 = \Sigma_2 \cup \Sigma_0 \cup J_1$

(iii) $J_0 \cap J_1 = I_0$ (a single point)

(iv)  $T$ has two inverse branches $T_\epsilon^{-1}$, $\epsilon = \pm$.

Another definition is

$$A_\epsilon = \{M \in \mathbf{R}^2;\ \epsilon J(M) < 0\}, \tag{5}$$

which implies

$$\mathbf{R}^2 = \overline{A}_+ \cup A_-, \quad A_\epsilon = T_\epsilon^{-1}(\Sigma_2).$$

As an example of such a situation, consider the two parameter family

$$T:\ (x,y) \to (y, hy - cx + x^2),\ \ 0 \leq h < 3/2,\ \ c \geq 1, \tag{6}$$

Its two inverse branches are

$$T_\epsilon^{-1}:\ (x,y) \to (\frac{c}{2} + \frac{\epsilon}{2}[c^2 + 4(y - hx)]^{\frac{1}{2}}, x),\ \ \epsilon = \pm, \tag{7}$$

and the two critical curves $J_0, J_1$ are

$$J_0:\ x = \frac{c}{2};\ \ J_1:\ y = hx - \frac{c^2}{4}. \tag{8}$$

The unstable fixed point $S(1 + c - h, 1 + c - h)$ of (6) is always located in $\Sigma_2 \cap A_+$; the focus $0(0,0)$ is always located in $\Sigma_2 \cap A_-$ and it destabilizes ($\forall h \in [0, 3/2[$) for $c = c_0 \equiv 1$, producing a singularity (an isolated closed invariant curve $\mathbf{C}$ or a pair of cycles $C_k$ stable node-saddle of order $k$, depending on the value of $h$), which at least in the neighbourhood of $c_0$, will also belong to $\Sigma_2 \cap A_-$.

For $h = 0$ ((6) is then a symmetrically decoupled mapping, cf. [9]) one obtains a pair $C_4$ of cycles: stable node-saddles of order $k = 4$ (the linear part $L$ of (6) for $c = c_0$ is then such that $L^4 = Id$);

this situation exists in a small neighbourhood of $h=0$. There exists also a value $\bar{h}$ $(\bar{h}\gg0)$ such that, for any $h \in ]\bar{h},3/2[$, the singularity produced by the destabilization of 0 is an isolated closed invariant curve **C**.

For a given value of $c$ the fixed point $S$ of (6) will be a saddle (of type 2) if

$$\frac{c+1}{3} \le h < \frac{3}{2};$$ (9)

for other values of $h$ $(0 \le h < (c+1)/3)$, $S$ will be an unstable node of type 2 (star node for $h=0$).

For the values of $h$ satisfying (9), the focus 0 thus produces for $c = c_0$ an invariant curve **C** which is the asymptotic limit of the continuation $\mathbf{B}_1^+$ of the consequent arc starting at the saddle $S$ ($\mathbf{B}_1^+$ winding as a spiral around **C**, see Fig. 1 and 5). The existence of **C** is thus assured for all values of $c$ associated with this situation. For other values of $h$ $(0 \le h < (c+1)/3)$, a similar qualitative situation may exist: the continuation of an invariant arc starting at an unstable node $S$ approaches asymptotically a winding around the set of four pairs of node-saddle points of $C_4$, and the singular closed invariant curve formed by the arcs $\mathbf{B}_4^+$ plays the role of **C**.

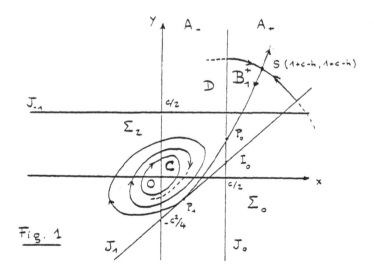

Fig. 1

The limiting situation ($h{=}0$) has analytical properties specific to the then decoupled form of (6) (cf. [9]): one knows analytically a countable infinity of submanifolds invariant under $T$ which pass through $S$   ($\{x{=}f_p(y)\} \cup \{y{=}f_{p+1}(x)\}$, $p \in \mathbb{N}, f(x) \equiv x^2 - cx$) and it is also possible to construct (by adapting the results of [2] to star nodes) the continuation of all invariant curves (of any slope) starting at $S$ (Fig. 2). In the case when the invariant curve $\mathbf{C}$ exists

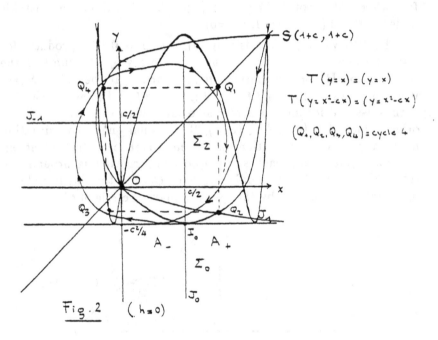

Fig. 2    ($h \neq 0$)

($h{\in}]\bar{h},3/2[$), it is possible to define a value $c_h^*$ of the parameter $c$ corresponding to the tangency of $\mathbf{C}$ and $J_0$ at a point $M_0$ (and necessarily, due to the assumptions (i) to (iv), to the tangency of $\mathbf{C}$ and $J_p$ at the consequent point $M_p = T^p M_0$ for all $p \in \mathbb{N}$) such that

$$\mathbf{C} \in \Sigma_2 \cap A_-, \quad \forall c \in ]c_0, c_h^*[.$$

Under these conditions it is easily shown that $\mathbf{C}$ is then necessarily globally invariant under $T$ as well as under $T_-^{-1}$:

$$TC = T_-^{-1}C = C, \quad \forall c \in ]c_0, c_h^*[. \tag{10}$$

A numerical algorithm using the property (10) permits us to obtain $c_h^*$: it is possible to show, for example ([4b]), that $c_1^* = 1.18735$.

One observes that, in the interval $]c_0, c_h^*]$ for increasing $c$, one of the following is necessarily true. The curve $\mathbf{B}_I^+$ tends asymptotically to $\mathbf{C}$ without self-intersection, without contact with $\mathbf{C}$, and with a single intersection with $J_0$ $(\mathbf{B}_I^+ \cap J_0 = P_0)$ (Fig. 1 and 5); or $\mathbf{B}_I^+$ tends to $\mathbf{C}$ with self-intersections due to several common points with $J_0$ (there exists in particular a value $\bar{c}$ for which $\mathbf{B}_I^+$ is tangent to $J_0$ at one point); finally $\mathbf{B}_I^+$ tends to $\mathbf{C}$ with the appearance of heteroclinic points (transverse intersections of $\mathbf{B}_I^+$ and $\mathbf{C}$) (Fig. 3).

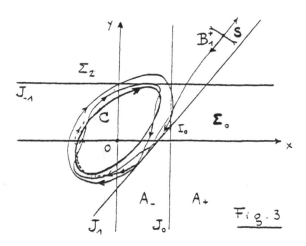

Fig. 3

In this interval $]c_0, c_h^*[$, the stable submanifold $\mathbf{C}$ has images under $T_\epsilon^{-1}$ without any common point

$$T_-^1\mathbf{C} = \mathbf{C}, \quad T_+^1\mathbf{C} = \mathbf{S}(\mathbf{C}), \tag{11}$$

$\mathbf{S}(\ )$ being the symmetry with respect to $J_0$.

For $c > c_h^*$, if one still has

$$T_-^1\mathbf{C} \cup T_+^1\mathbf{C} = \mathbf{C} \cup \mathbf{S}(\mathbf{C}), \tag{12}$$

then $\mathbf{C}$ is no longer contained in $\Sigma_2 \cap A_-$ and is no longer globally invariant under $T^{-1}$, because then it is possible to show that

$$T_-^1\mathbf{C} = \overline{\mathbf{C}}_- \cup \mathbf{S}(\mathbf{C}_+), \tag{13}$$

where by definition

$$\mathbf{C}_\epsilon = \mathbf{C} \cap A_\epsilon. \tag{14}$$

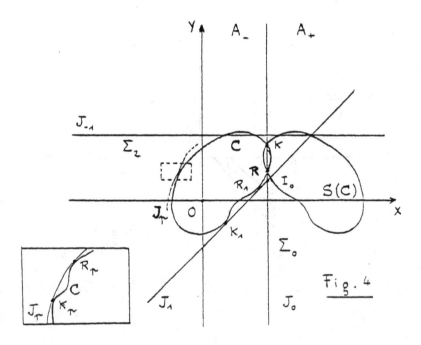

Fig. 4

The above imply essentially that, if the curve **C** is $C^1$, it will exhibit for $c \geq c_h^*$ an infinity of convexity changes at consequent points of $\overline{C}_+ \cap J_0$, which are osculation points, at least of order one, of **C** and of the critical curves $J_p$ of $T$ $(p=1,2,....)$ (see Fig. 4).

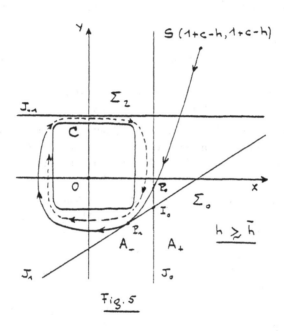

Fig. 5

Dynamic Systems Research Group,
Universite' Paul Sabatier
118, route de Narbonne
F-31062 Toulouse
France

## 4. References

[1]   Birkhoff, G.D.: 1950, *Collected mathematical papers.* Vol. 1 and 2, AMS
       Publications, New York.

[2]   Lattès, S.: 1906, Sur les équations fonctionnelles qui définissent une courbe
       ou une surface invariante par une transformation. *Mat. Pura Appl.* (3) *13*,
       1-37.

[3]   Gumowski, I. and C. Mira: 1980, *Recurrences and discrete dynamic
       systems.* Lecture Notes in Math. 809, Springer Verlag, Berlin, Heidelberg,
       New York.

[4]   Clerc, R.L. and C. Hartmann: (a): 1979, Quelques propriétés de certaines
       singularités unidimensionnelles d'une récurrence autonome d'ordre deux.
       *C.R. Acad. Sci. Paris*, Sér. A. *289*, 31-34. (b): 1980, Sur les bifurcations
       d'une courbe invariante isolée fermée d'une récurrence du second ordre.
       *C.R. Acad. Sci. Paris*, Sér. A. *290*, 1005-1008.

[5]   Clerc, R.L. and C. Hartmann: 1981, Numerical search for invariants of a
       discrete dynamic system. In: *Nonlinear problems of mechanics in geometry
       and analysis* (Proc. Symp., Univ. Paul Sabatier, Toulouse, 1979). Res. Notes
       in Math., 46, Pitman, Boston, Mass., pp. 93-101.

[6]   Clerc, R.L. and C. Hartmann: 1981, Antecedent invariant curves of an
       endomorphism. Influence domain of a stable cycle coexisting with an
       isolated stable invariant curve. *Springer series in* SYNERGETICS *9*, 47-53.

[7]   Kawakami, H. and K. Kobayashi: 1979, Computer experiments on chaotic
       solutions of $X(t+2)-aX(t+1)-X^2(t) = 0$. *Bull. Fac. Engin. Tokushima
       Univ.* 16, 29-46.

[8]   Rogers, T.D. and B.L. Clarke: 1981, A continuous planar map with many
       periodic points. *Appl. Math. Comput.* 8, 17-33.

[9]   Clerc, R.L. and C. Hartmann: 1981, Invariant manifolds of separable
       discrete dynamic systems. *Preprint Univ. Paul Sabatier,* Toulouse III, no.
       81-2, Presented at "Dynamic Days", La Jolla, California, 1982.

A. Sklar

## The cycle theorem for flows

A family $\{f_t | t \text{ in } \mathbf{I}\}$ of functions is a *flow* if the following conditions are satisfied.

1. The index set $\mathbf{I}$ is a set of real numbers that includes the number 1, is closed under addition ($s,t$ in $\mathbf{I}$ implies $s+t$ in $\mathbf{I}$), and is 'half-closed' under subtraction ($s,t$ in $\mathbf{I}$ implies $|s-t|$ in $\mathbf{I}$).

2. For any $s,t$ in $\mathbf{I}$,

$$f_{s+t} = f_s \circ f_t, \tag{1}$$

where $\circ$ indicates *composition*.

It follows from (1) that for any $s$ in $\mathbf{I}$ and any positive integer $m$,

$$f_{ms} = f_s^m, \tag{2}$$

where $f_s^m$ is the $m$th *iterate* of $f_s$. Hence $f_s$ is an *iterative root of order* $m$ of $f_{ms}$. Calling a flow *divisible* if $t$ in $\mathbf{I}$ implies $t/n$ in $\mathbf{I}$ for every positive integer $n$, it follows that every function in a divisible flow has iterative roots of all orders.

An *n-cycle* of a function $f$ is a set of $n$ *distinct* points $x_0, x_1, \ldots, x_{n-1}$ in the domain of $f$ such that

$$f(x_m) = x_{m+1} \text{ for } 0 \leq m \leq n-2, \quad f(x_{n-1}) = x_0. \tag{3}$$

The purpose of this paper is to prove the following.

CYCLE THEOREM FOR FLOWS. Let $\{f_t | t \text{ in } \mathbf{I}\}$ be a divisible flow. Suppose there is a $t > 0$ in $\mathbf{I}$ and an element $x_0$ such that

*J. Aczél (ed.), Functional Equations: History, Applications and Theory, 227-229.*

$$f_t(x_0) = x_0 \text{ but } f_{rt}(x_0) \neq x_0$$

for any rational $r$ in $(0,1)$.                                     (4)

Let $m,n$ be relatively prime positive integers. Then $f_{(m/n)t}$ has infinitely many $n$-cycles.

Proof. For $n \geq 2$ and $m$ any positive integer $\leq n-1$, set

$$x_m = f_{(m/n)t}(x_0).$$                                     (5)

It follows from (1) that $f_{t/n}(x_m) = x_{m+1}$ for $0 \leq m \leq n-2$, and from (4) and (1) that $f_{t/n}(x_{n-1}) = x_0$. Hence the elements $x_0, x_1, \ldots, x_{n-1}$, if *distinct*, form an $n$-cycle of $f_{t/n}$. Now suppose that the elements $x_0, \ldots, x_{n-1}$ are not all distinct. This means that there are integers $l, m$, with $0 \leq l < m \leq n-1$, such that $x_l = x_m$. Now we cannot have $l=0$, for that would mean

$$x_0 = f_{(m/n)t}(x_0),$$

contrary to (4), since $m/n$ is a rational number in $(0,1)$. So $1 \leq l < m \leq n-1$, whence by (5),

$$f_{(l/n)t}(x_0) = f_{(m/n)t}(x_0).$$

Using this, together with (4) and (1), we obtain

$$x_0 = f_t(x_0) = f_{(n/n)t}(x_0) = f_{((n-m)/n)t + (m/n)t}(x_0)$$
$$= f_{((n-m)/n)t}(f_{(m/n)t}(x_0)) = f_{((n-m)/n)t}(f_{(l/n)t}(x_0)),$$

whence

$$x_0 = f_{((n-m+l)/n)t}(x_0).$$

But $(n-m+l)/n$ is a rational number in $(0,1)$, so the last equality contradicts (4). So the elements $x_0, x_1, \ldots, x_{n-1}$ are all distinct, and thus form an $n$-cycle of $f_{t/n}$.

To complete the argument, let $n$ be a positive integer, and $m$ an integer $\geq 2$. Then, as has just been shown, $f_{t/(mn)}$ has (at least) one $mn$-cycle. Now it is virtually immediate that, if a function $f$ has an $mn$-cycle, then under iteration of $f$, this cycle splits into $m$

$n$-cycles of $f^m$. Hence $f^m_{t/(mn)}$ has (at least) $m$ $n$-cycles. but by (2), $f^m_{t/(mn)} = f_{t/n}$, so it follows that $f_{t/n}$ has (at least) $m$ $n$-cycles *for every* $m \geq 2$. Hence $f_{t/n}$ has infinitely many $n$-cycles.

Finally, (2) yields

$$f^m_{t/n} = f_{(m/n)t}$$

for any positive integers $m, n$. If $m$ and $n$ are relatively prime, then it is easy to see that any $n$-cycle of $f_{t/n}$ remains an $n$-cycle of $f^m_{t/n}$. Thus $f_{(m/n)t}$ has infinitely many $n$-cycles, and the proof of the theorem is complete.

As an example of the Cycle Theorem, consider the flow $\{f_t | t \text{ in } \mathbf{R}\}$ in which, for any $t$ in $\mathbf{R}$, the function $f_t$ is defined on the *projective line* (the real line $\mathbf{R}$ augmented by a single $\infty$) by

$$f_t(x) = \frac{x \cos \pi t - \sin \pi t}{x \sin \pi t + \cos \pi t} \, .$$

Hence $f_1(0)=0$, while, for $t$ in $(0,1)$, $f_t(0) = -\tan\pi t \neq 0$ (n.b. $f_{1/2}(0)=\infty$). Therefore the Cycle Theorem yields the fact that, for $m, n$ relatively prime positive integers, $f_{m/n}$ has infinitely many $n$-cycles. In fact, for this particular flow, it can be shown that, for $m, n$ relatively prime integers with $n > 0$, *every* point on the projective line is in an $n$-cycle of $f_{m/n}$, while, if $t$ is irrational, then *no* point on the projective line is in a cycle, of any order, of $f_t$.

Finally, it may be noted that the condition $f_{rt}(x_0) \neq x_0$ *for any rational* $r$ *in* $(0,1)$ in (4) can be replaced by the simpler condition $f_{t/n}(x_0) \neq x_0$ *for all integer* $n \geq 2$.

Department of Mathematics
Illinois Institute of Technology
Chicago, IL 60616
U.S.A.

# INDEX

# Mathematics and Its Applications

Managing Editor:

M. HAZEWINKEL
*Centre for Mathematics and Computer Science, Amsterdam, The Netherlands*

1. Willem Kuyk, *Complementarity in Mathematics, A First Introduction to the Foundations of Mathematics and Its History.* 1977.
2. Peter H. Sellars, *Combinatorial Complexes, A Mathematical Theory of Algorithms.* 1979.
3. Jacques Chaillou, *Hyperbolic Differential Polynomials and Their Singular Perturbations.* 1979.
4. Svtopluk Fučik, *Solvability of Nonlinear Equations and Boundary Value Problems.* 1980.
5. Willard L. Miranker, *Numerical Methods for Stiff Equations and Singular Perturbation Problems.* 1980.
6. P. M. Cohn, *Universal Algebra.* 1981.
7. Vasile I. Istrăţescu, *Fixed Point Theory, An Introduction.* 1981.
8. Norman E. Hurt, *Geometric Quantization in Action.* 1982.
9. Peter M. Alberti and Armin Uhlmann, *Stochasticity and Partial Order. Doubly Stochastic Maps and Unitary Mixing.* 1982.
10. F. Langouche, D. Roekaerts, and E. Tirapegui, *Functional Integration and Semiclassical Expansions.* 1982.
− C. P. Bruter, A. Aragnol, and A. Lichnerowicz, *Bifurcation Theory, Mechanics and Physics.* 1983.